A THEORY FOR EVERY= THING

Jeremy Bernstein

A THEORY FOR EVERY=THING

COPERNICUS
AN IMPRINT OF SPRINGER-VERLAG

Some of the pieces appearing in this book were published—often in substantially different form—in the following publications: "Language" in *The Atlantic Monthly*. "The Drawing" as "What Did Heisenberg Tell Bohr About the Bomb?" and a "A Brief History of Black Holes" as "The Reluctant Father of Black Holes" in *Scientific American*. Revised and reprinted with permission. Copyright © 1995 and 1996, respectively, by Scientific American, Inc. All rights reserved. "Bohr" as "King of the Quantum"; Segrè" as "Eye on the Prize"; "Madame Curie" as "The Passions of Madame Curie"; and "Linus Pauling" as "Odd Man In" in *The New York Review of Books*. Revised and reprinted with permission. "The Philosophy Circle" and "The Faculty Meeting" first appeared in slightly revised form in *The New Yorker*, as did "Smuggler" in a substantially revised form as "Report from Aspen." Revised and reprinted with permission. "Julian Schwinger" and "Portrait of Bleibermacher" first appeared in the *American Scholar*, the latter in a substantially revised form. Copyrighted and used by permission of the author. Grateful acknowledgment is also made to the Niels Bohr Archive for permission to reprint from a letter from Bohr to James Chadwick on pp. 78–79.

All characters except for historical personages are fictitious.

Published in the United States by Copernicus, an imprint of Springer-Verlag New York, Inc.

Copernicus
Springer-Verlag New York, Inc.
175 Fifth Avenue
New York, NY 10010

Library of Congress Cataloging-in-Publication Data

Bernstein, Jeremy, 1929–
    A theory for everything / Jeremy Bernstein
        p.   cm.
    Includes bibliographical references and index.
    ISBN 0-387-94700-0 (hardcover : alk. paper)
    1. Physics—History.   2. Physicists—Biography.   I. Title.
QC7.B49   1996
530—dc20   96-15531

Manufactured in the United States of America.
Printed on acid-free paper.
Designed by Irmgard Lochner.

9 8 7 6 5 4 3 2 1

ISBN 0-387-94700-0   SPIN 10529496

Contents

Preface

It is customary—indeed I have done it myself more than once—for the author of a collection like this to write a long preface explaining why all the pieces of writing that have been fitted together under one roof belong there. In general, the explanation is quite simple and boils down to two propositions: First, there is a commonality of subject matter, and second, there is a commonality of authors—namely, the author in question. (My friend and colleague John McPhee—and this took some nerve—actually published a collection of diverse pieces under the title *Table of Contents*. The book's jacket was the table of contents.) Most of these prefaces are rather tedious.

On the other hand, I have always found it quite interesting when the author of one of these collections steps outside an essay to comment on what has happened since he or she wrote it: "They all didn't live so happily after all." That is what I have done in this collection in a series of postscripts. When only a little retouching was required to bring an article up to date, I have done that in the body of the article. But in many of the pieces, a coherent unit would have become unstuck had I intervened before the end. I have also written some "prescripts," which explain how some of these articles got written in the first place. In particular, in the section entitled "Leptons"—lighter pieces—I describe my brief adventure with the *New Yorker* magazine's fiction department. Incidentally, although I wrote for that magazine for some 30 years, only three of the pieces in this collection first saw the light of day there. Some have never been published before.

It is also customary in these prefaces to thank one's agent and one's next of kin. My agent, Elsie Stern, died a few years ago before I had the chance. Let me thank her here. Also I am not sure if I ever properly thanked William Shawn before he died. Those of us who had the chance to work with him had an experience that will be with us forever. I *can* thank Bill Frucht and Jerry Lyons of Copernicus for nurturing this book. I will thank my next of kin on Sundays and holidays.

Jeremy Bernstein
New York City
August 1996

A
THEORY
FOR
EVERY=
THING

Newton's Apple and Einstein's Elevator

When Charles Montague (Lord Halifax) died in 1715, he left the sum of £100 to Isaac Newton "as a Mark of the great Honour and esteem I have for so Great a Man." But in the same will, he bequeathed the sum of £5,000 to Newton's niece, Catherine Barton, along with a magnificent house and the income to maintain it. While Halifax may have honored and esteemed Miss Barton, it was widely assumed that, in the words of Dryden's poem:

> At Barton's feet the God of Love
> This Arrow & his Quiver lays. . .

A curious feature of this arrangement was that Newton certainly must have known about it, since he and his niece, popularly known as La

Bartica, were living under the same roof. In sexual matters Newton was totally puritanical. According to his doctor, Richard Mead, he died a virgin. Mead told this to Voltaire who then told it to everyone. Why Newton countenanced this arrangement between Halifax and La Bartica is a question historians still debate. Some contemporary gossip had it that Newton used this liaison to obtain his position as Master and Worker of the Mint. Halifax had become First Lord of the Treasury in 1697 and had appointed Newton the year before, when he was Chancellor of the Exchequer. It was Newton, incidentally, who invented the idea of milling the edges of coins to keep them from being so easily counterfeited.

In 1717, when she was 38, La Bartica married John Conduitt, a wealthy man some nine years her junior. Newton was then 75. He lived another ten years, during which period Conduitt spent a great deal of time extracting biographical information from him—or what was alleged to be biographical information—which he recorded in memoranda. In one of these memoranda Conduitt presented the story of Newton's apple. It begins "In the year 1666 he [Newton] retired again from Cambridge [1666 is often referred to as the Annus Mirabilis—the year of wonders—the year Newton created classical physics and the mathematics needed to express it. It was also a year in which the presence of the plague caused him to flee Cambridge temporarily]. . . . to his mother in Lincolnshire & and whilst he was musing in a garden it came into his thought that the power of gravity (which brought an apple from the tree to the ground) was not limited to a certain distance from the earth but that this power must extend much farther than was usually thought."

This is Newton's apple. We have all heard of it. I first read about it, when I was still a child, in a book called *Stories from British History*. It had a charming picture of Newton sitting in a garden and staring at an apple tree. Neither then nor for years afterwards did I have the remotest idea why this story was interesting. What on earth could a falling apple have meant to Newton that it didn't mean to everyone else in Lincolnshire, who must have watched apples fall from trees on a daily basis? Indeed, it was not until the spring of 1948, my freshman year at Harvard, that I understood the meaning of this story and had my first inkling of what deep science was all about.

When I entered Harvard, in the fall of 1947, I had no idea what I wanted to become, except that I was certain it would have nothing to do with science. I had taken one science course in high school—physics—and found it impossibly dull. I could not imagine that a serious adult, Einstein notwithstanding, would spend his or her life doing this sort of thing for a living. I thought for awhile—God knows why— that I might go into economics, but I was spared this fate by the Harvard General Education program.

After the war, James Bryant Conant, the president of Harvard University, decided that the traditional methods of teaching science to nonscience majors were not good enough. Harvard had a science requirement which, like the requirement that every undergraduate be able to swim two laps in the gymnasium pool, had to be filled by one's senior year. The traditional way of doing this was to take a watered down version of the standard introductory science course in a science of one's choice. "Rocks for Jocks" (geology) was very popular and I think one could also satisfy the science requirement by taking a course in celestial navigation that would enable the more affluent undergraduates to pilot the family yacht from Newport to Bermuda.

Conant, who had been deeply involved in the organization of wartime science and was himself a chemist, thought Harvard College could do better than this, so he set about creating a General Education program that would become the freshman core curriculum. Within this program there were so-called Natural Sciences courses designed for nonscientists like myself. To help steer one through this maze of choices was the Harvard Confidential Guide to Freshman Course [sic], an irreverent course guide that tipped one off to the easiest courses and most sympathetic instructors. On this basis I selected Natural Science 3, taught by the historian of science I. Bernard Cohen. In later years, for some reason, Cohen took a drubbing from the Confy Guide,* but for a total cipher in science, as I was then, the course was just fine and Cohen a good instructor.

In any event, whatever the merits of Cohen's course, I am eternally grateful to him for guiding me to my first great teacher in physics, Philipp Frank. It was Professor Frank who taught me the significance of Newton's apple. I came upon him because of an interest I had developed in Einstein's theory of relativity, of which I understood next to nothing. That was the problem, and Bernard Cohen thought Frank might be the solution. Frank was then teaching a course called Physics 16, a kind of introductory course in modern physics with an emphasis on its philosophy and history. It promised to explain relativity on an elementary basis so I enrolled even though I didn't quite have the prerequisites. The course met once a week, on Wednesday afternoons, in the old Jefferson physics laboratories. On the first Wednes-

*Some years later, when I was earning my way through graduate school by teaching sections in the more rigorous version of this course, Natural Science 2, I was also evaluated in the Confy Guide. "Bernstein," the guide said, "handled the bulk of the section work last year. He did not exactly provide a 90-minute fount of inspiration every week, but this is probably because he was as bored as everyone else with the useless sections. For his part he was always prepared for the required repetitive lecturette, if anyone cared to stay around for it after the quiz. He was also willing to help students individually with difficult material, and he added a bit of whimsy to the course. . . ." The anonymous writer of this screed probably didn't realize that poor Bernstein was working about a hundred hours a week on his thesis and what the writer perceived as whimsy was actually bug-eyed desperation brought about by lack of sleep.

day of the spring semester I was in place, ready to master the theory of relativity.

Cohen had told me a few things about Professor Frank. I knew he was Viennese. I later learned that he had been born in that city in 1884, which means that when I enrolled in his course he was about the same age I am now. I also knew that he had just written a biography of Einstein—*Einstein: His Life and Times*—which I went out and bought immediately. I later learned that he had been in personal and professional contact with Einstein since 1907, when Frank had written a philosophical paper that Einstein read and criticized. Five years later, Einstein had recommended him as his successor at the German University in Prague, where Frank remained until 1938. At this time, Harvard cobbled together some kind of half-time appointment, not quite in the physics department and not quite in the philosophy department. Hence this course, which was somewhere between the two disciplines.

I will never forget the initial impression Professor Frank made on me. He was a short, ovoid man who walked with a pronounced limp, apparently the result of a childhood encounter with a Viennese streetcar. His head was largely untroubled by the presence of hair, and the hair that remained seemed to lead a life of its own, sprouting in random directions. He looked like an extremely intelligent elderly basset hound. He had an accent in English that was difficult to place. I used to do imitations of it with some success. I finally realized that it reflected the seven or eight languages that he had spoken in Europe. They were piled on one another like the cities of Troy, with English near the top. He seemed to have read almost everything in these languages as well as in others such as Arabic, Greek, Hebrew and Persian. He was, one soon realized, incredibly funny in that earthy, cynical Viennese way. Ultimately, I spent the next ten years at Harvard and came to know him pretty well. When he died in 1966 I spoke at his memorial service.

The course began in what at first seemed to me an odd place, although later I understood what he had in mind. We started by examining the nature of mathematical truth, with Euclidean geometry as our example. One can consider Euclidean geometry as an empirical science. "Geometry," after all, means "earth measure." From this point of view, the proposition that a triangle has three interior angles whose sum is 180 degrees becomes an experimental question to be answered by measuring the angles of any given triangle. Indeed, if one draws such a triangle on the surface of the earth the angle sum will not be 180 degrees, but larger, since the earth is a sphere. Its surface, we would say, has a positive curvature and geometry done on it is non-Euclidean.

On the other hand, we can also regard plane geometry as a deductive science. Starting from Euclid's postulates, we can prove that anything having the attributes of a triangle must have an interior angle sum of exactly 180 degrees. This is a logical truth but it might not have any empirical content. There might not be any object in the universe made out of sticks and stones, or whatever, that had this property. All real objects might be non-Euclidean. Insofar as mathematics is perfectly true, it might be perfectly empty.

That lesson having been absorbed, Professor Frank went on to give a somewhat impressionistic account of the history of the science of mechanics—how things move. The Greeks, especially Aristotle and Plato, had a teleological view of motion. Earth, air, fire and water moved the way they did because they were earth, air, fire and water and simply were following their natures. The Moon and planets appeared to have more perfect regular motions so they were assigned a different essence—the quintessence, or fifth essence. In our language, the laws of motion were not, in this cosmology, "universal." Different laws of motion applied to different essences. This view was commonly held right up to Newton's time. John Donne, who died in 1631, eleven years before Newton was born, reflects it in his poem "A Valediction: forbidding mourning." He writes:

> Moving of th' earth brings harmes and feares.
>> Men reckon what it did and meant,
> But trepidation of the spheares,
>> Though greater farre, is innocent.

Galileo, who died the year Newton was born, found himself in terrible trouble with the Establishment in part because he claimed that the "trepidation" of the celestial spheres was no more innocent than any other form of motion. Indeed, one of the things Galileo did with his telescope was to examine the Moon and discover that it had mountains and a surface that appeared to be as corrupt as the Earth's.

Newton took this one step farther—the decisive step—and this is where the apple comes in. Imagine, Professor Frank told us, a thought experiment—a *gedanken* experiment—in which the apple tree slowly grows to incredible heights. If we stop the tree at some height and let the apple fall, there is no reason to think that it will not fall to the Earth. The force of gravity will no doubt grow weaker on the apple, but why should it vanish altogether? Now let the tree become as tall as the distance between the Earth and the Moon and let the Moon replace the apple on the tree. But then the Moon must be falling towards the Earth under the influence of gravity! That is what Newton's apple is about—the universality of gravitation. Indeed, Conduitt goes on to say just this. He summarizes Newton's reasoning inspired by the falling apple:

"Why not as high as the moon [the apple tree] said he to himself & and if so that must influence her motion & perhaps retain her in her orbit, whereupon he fell a calculating what would be the effect of that supposition. . . ."

It would both trivialize the matter and be historically wrong to say that this flash of insight led Newton immediately to his quantitative theory of gravitation and mechanics any more than, as we shall see, Einstein's elevator led him immediately to his general theory of relativity and gravitation. But in both cases, these were liberating insights that revealed unexpected depths in commonplace experiences. It took Newton 20 years before his *Principia,* the masterpiece in which he presented the full dynamic scheme, was finally published in 1686. The intellectual obstacles he overcame are almost beyond understanding. He had to invent both differential and integral calculus, for example, and it took him over a decade to understand the exact mathematical form of gravitational force. It is one thing to say that the force falls off somehow as the distance between two massive objects increases. It is quite another to realize, as Newton eventually did, that a force law in which the decrease is precisely as the square of the distance produces orbits that are conic sections—circles, ellipses, parabolas and hyperbolas; the planets move in ellipses—to say nothing of the idea that it is the balance of forces that determines accelerations: the law $F = ma$. All of this we now teach in a few weeks to our beginning physics students, who simply take it for granted. I am glad I began my studies in the odd way that I did, so that I learned from the beginning just how hard this all had been.

This is science at its deepest level. It can, I believe, be taught to almost anyone—certainly any college student. Whether it *should* be taught to everyone is another question. Should there be a science requirement any more than a requirement that every student take, say, a year of Chinese or learn to play a musical instrument? I am not sure why Harvard had a science requirement in the first place, at least before the war. But in 1947, Conant published a book called *On the Understanding of Science* in which he gave some of the reasons he considered the teaching of science so important in the nuclear age. The same reasons are usually given now. Under the heading "Science and the National Policy" Conant writes,

> In a democracy, political power is widely diffused. National policy is determined by the interaction of forces generated and guided by hundreds of thousands if not millions of local leaders and men of influence. [These were the days when Harvard College was a single-sex institution and Radcliffe girls—they were called girls—hostessed dances called "Jolly-ups."] Eventually within the limits imposed by public opinion, decisions of far-reaching importance are made by a relatively few. These men are almost accidentally thrown into positions of temporary power by the forces working throughout our benignly chaotic system of political democracy. Because of the fact that the applications of science play so important a part in our daily lives, matters of public policy are profoundly influenced by highly technical scientific considerations. Some understanding of science by those in positions of authority and responsibility as well as by those who shape opinion is therefore of importance for the national welfare.

This is certainly true, but the courses that were created to meet this supposed need were irrelevant if not absurd. How is a one year "Physics for Poets" or a Natural Science 3 course going to equip anyone to deal with "highly technical scientific considerations"? We saw this in the debate over nuclear power. Experts who had spent lifetimes studying the safety of nuclear reactors could not agree on how safe the reactors were. How is a course that might spend a week studying the atomic nucleus on the most superficial level going to prepare anyone for this debate?

Mind you, I think teaching science to nonscientists is a good thing. I have been doing it for much of my life, but the suggestion that taking a year of popular science will enable one to contribute on an informed basis to these debates seems nonsensical to me. If we decided, for example, that it was vital to our political process that the citizenry learn Japanese, we would not attempt to meet this need by creating a one-year "Japanese for Poets" course of which three quarters dealt with the history of the language. We would immerse ourselves in Japanese day and night, which was how they used to teach languages in the army.

I am equally unimpressed by C.P. Snow's discussion of "the two cultures." I wonder if anyone still reads him. Some of the novels—*The Masters* or *The Light and the Dark*—are quite good. One curious feature, however, is that his scientists, even the articulate ones, never talk about their science. Some never talk at all. In *The Light and the Dark*, Snow has the Master of his fictional college say "Isn't that just the trouble with some of your scientific colleagues, Mr. Getliffe. . . .They're always saying the last word, but they never seem to say the first." Indeed, what Snow is worried about most when it comes to the two cultures seems to be High Table conversation. I remind you of what he actually said in his 1959 Rede Lecture where this was first discussed. He remarked that he has been among nonscientists who, from time to time, gave "a pitying chuckle at the news of scientists who have never read a major work of English literature. Once or twice I have been provoked and have asked the company how many of them could describe the Second Law of Thermodynamics. The response was cold: It was also negative." I have never understood why it was so important that a Cambridge classics don be able to describe the Second Law of Thermodynamics or, indeed, why it mattered in the grand scheme of things if some scientist was semiliterate. The loss was theirs. What is important is that the cultural opportunity be made available. To do this it may be necessary to have required courses. I am a perfect example. Had there been no science requirement at Harvard I would have taken no science. I would never have encountered Professor Frank and never have become a physicist. I am immensely grateful that there was such a requirement. It transformed my life.

Newtonian physics transformed the entire intellectual life of Europe for the next 200 years. It is easy to say that the reason was the success of Newtonian physics, but this does not do him full justice. Newton changed the meaning of what a successful physical theory was. Take the difference between Newton and Kepler. Kepler died only twelve years before Newton was born, but his science was of an entirely different age. After many years of excruciating calculation, Kepler realized that the orbit of Mars was an ellipse. This was a great empirical discovery, but Kepler could give no reason why it was true. From his point of view, the orbit might have been any closed curve. Newton changed all this. The planetary orbits had to be elliptical because the law of force involved the inverse square of the distance between the gravitating masses. This firm prediction of Newton's theory of universal gravitation also applied to the Moon and the comets as well as the planets. Newton revealed the almost incredible predictive power of mathematical physics, which led to a kind of simplistic determinism. Given the forces and the initial conditions of every atom—at least in principle—the entire past and future of the universe appeared to be determined. There seemed to be no place for free will. The universe, according to this cosmology, is a kind of relentless machine. Quantum mechanics, with its uncertainties and probabilities, has redefined this question.

Despite the advent of relativity and the quantum theory, our everyday language remains largely Newtonian. Take Conant's phrase quoted above: "National policy is determined by the interaction of forces. . . ." The idea of something being determined by the interaction—the balancing—of forces is purely Newtonian. We think this way without thinking twice about it. Newtonian mechanics was so successful that I am not aware of any serious challenge to it until the end of the nineteenth century—indeed, until 1883, when Ernst Mach published his polemical work *The Science of Mechanics*.

That this book had such an influence on people like Einstein is somewhat odd. Born in 1838 in the Austro-Hungarian town of Chirlitz, Mach was not a very distinguished physicist. The only work in physics for which he is remembered is on supersonic motion—the Mach numbers. When he wrote his book he was teaching at the German University in Prague, where both Einstein and Professor Frank would eventually land. Sometime around 1913, Professor Frank orchestrated a meeting between Mach and Einstein in Vienna to discuss atoms. Mach didn't believe in them and Einstein did. Mach's favorite question was "Have you seen one?" As usually happens in these matters, with the success of the atomic theory Mach's largely philosophical objections were forgotten. This aside, the strange thing about Mach's seminal book is that it was not written in response to any new experimental data. He states things baldly in the preface to the first edition, where he writes "The present volume is not a treatise upon the application of the principles of mechanics. Its aim is to clear up ideas, expose the real significance of the matter, and get rid of metaphysical obscurities." What obscurities? To understand what Mach had in mind, let us return to Newton's apple.

Newton's thought experiment suggested to him that the Moon was being pulled towards the Earth by the force of gravity. But what holds it up? What force balances gravity to keep the Moon from falling down? What probably comes to mind is centrifugal force—the force that wants to push us off the merry-go-round. But this is a very peculiar force. It is not associated in any obvious way with material objects. A top spinning in empty space would, Newton would claim, bulge at its equator because the centrifugal force is at its maximum at the equator while vanishing at the poles. Furthermore, we can make the centrifugal force disappear by changing our point of view. The forces on an astronaut in a circular orbit cancel each other out as we view the situation from Earth. But to the astronaut there is no force; he or she is weightless.

In textbooks the centrifugal force is sometimes referred to as "fictitious" since it can be transformed away by changing the reference system. But this means Newton's law relating to forces and accelerations is ambiguous until we can determine our absolute state of motion. We don't know whether or not to include fictitious forces unless we know what reference system we are in. Newton must have recognized this ambiguity because he introduced the notion of absolute space, which he eventually identified with the sensorium of God. The omnipresence of God was Newton's ultimate reference system in which the force law was defined. Mach found this mixture of physics and metaphysics intolerable. He clearly saw what was needed—a theory without absolute space in which all the motions would be determined by the presence of massive objects. But he did not have the genius to create such a theory. That was left to Einstein.

Einstein's first insight came in 1907. By then, although he still had no academic job, his material condition had improved. He had received a promotion to Technical Expert Second Class in the Swiss

National Patent Office in Bern. He was earning 4,500 Swiss francs a year and was married and the father of a son. In 1905 he had had his own Annus Mirabilis, publishing the three papers that created modern physics. The paper on relativity came last and was preceded by the paper which created the quantum theory. It was for the work on the quantum that he received the Nobel Prize in 1921. In 1906, he published a short paper in which he derived the formula $E = mc^2$. A few of his contemporaries—not many—were beginning to get the idea that they had another Newton on their hands.

Some years later, Einstein recalled the precise moment in 1907 when he had the first flash of insight that ultimately led him to his scientific masterpiece, in 1915—his general theory of relativity and gravitation. "I was sitting in a chair," he reminisced, "in the patent office at Bern when all of a sudden a thought occurred to me: `If a person falls freely he will not feel his own weight.' I was startled. This simple thought made a deep impression on me. It impelled me towards a theory of gravitation." Einstein once called this the "happiest thought of his life." This was the Einstein elevator.

I am not sure when the actual image of an elevator came into Einstein's mind—whether it was in 1907 or later—but by 1916, when he wrote his wonderful popular book on relativity, simply entitled *Relativity*, he was using the image of an elevator, which he described as a "spacious chest resembling a room." This chest, he stipulates, has been placed in a portion of empty space "removed from stars and other appreciable masses." In it, there is an observer who "must fasten himself with strings to the floor, otherwise the slightest impact against the floor will cause him to rise slowly towards the ceiling of the room." Now Einstein imagines attaching a rope to a hook fixed on the lid of the chest and a "being"—"what kind of being is immaterial to us"—pulls on the rope with a constant force.

This means, according to Newton's law $F = ma$, that the chest (the elevator) will rise with a constant acceleration. In this new situation, the observer will sense the motion. Someone standing on the floor will feel the pressure, just as we do in a real elevator. If the observer drops something, it will appear to accelerate towards the floor, since the floor of the elevator is accelerating upward to meet it. All objects will appear to fall with the same acceleration, which, from this perspective, is the acceleration of the elevator floor towards the falling objects.

All of this seems obvious enough, almost trite. But the next step was the flash of genius, and like Newton's falling apple, its extraordinary implications are not at all evident. Einstein realized that the observer in the accelerating elevator can attach an entirely different meaning to the phenomenona he or she is experiencing. The observer could conclude that the elevator is at rest—not accelerating but suspended, unmoving, from the hook. But the elevator exists in a uniform field of gravitation, the kind we experience near the surface of the Earth. Since all objects in a uniform gravitational field fall with the same acceleration, this would explain why all the objects fall to the elevator floor with the same acceleration.* Thus, the Einstein elevator can be described in two perfectly equivalent ways—something that Einstein "elevated" to a general principle, the Principle of Equivalence. Either we can say there is no gravitational field but the elevator is uniformly

---

*This is what Galileo claimed to have demonstrated by dropping various weights from the Leaning Tower of Pisa. They all, he insisted, fell at the same accelerated rate. It is not clear whether Galileo ever actually performed this experiment and it is still less clear what he would have found if he had, since the air friction would have interfered with the gravitational effect. But it is a demonstration we do all the time in introductory physics classes. We take a penny and a feather and put them in a glass tube. The air is then pumped out, and lo and behold, the two objects fall at exactly the same rate—something that is predicted by Newtonian mechanics.

accelerating, or we can say that the elevator is at rest in a uniform gravitational field.

Einstein immediately saw the use for this idea, although he needed another eight years to create his general theory of relativity in its final form. But in 1907, he already saw that the Principle of Equivalence implied that gravitation altered the properties of space and time. It "curved" space, as we would put it, or more generally, it curved space-time. How does this come about? Using Einstein's model, we can imagine experimenting with light in our elevator. For example, suppose that inside our elevator, before it begins to accelerate, we have suspended some atoms that emit light of a definite color. We sit on the elevator floor and observe this light as it strikes our eyes. Now the "being" tugs on the rope and the floor of the elevator accelerates up to meet the light. But we know what happens when an observer and a light source are set in motion with respect to each other. There is a Doppler shift. In this case the light is shifted towards the blue since we are rushing upwards to meet it. But the Principle of Equivalence tells us that there will be exactly the same effect if the elevator is at rest in a gravitational field. Gravity alters the color of light—its frequency. But light frequency is a kind of clock, so Einstein would argue that gravity alters the rate of clocks. The theory also predicts how this rate will change in a given gravitational field. This prediction has been precisely tested using very accurate atomic clocks. The most extreme example of gravitational time change would be apparent if we could follow a clock as it entered a black hole. As it approached the boundary of the hole we would observe it going slower and slower. At the boundary it would stop altogether. Time would freeze.

Equally remarkable things happen to space. Imagine again that our elevator is at rest in empty space. Someone shines a light in at one side. Once inside, the light beam travels parallel to the floor then exits the other side. Now our "being" tugs on the elevator and we repeat the experiment. This time the elevator's upward motion causes the light beam to hit the other side closer to the floor. The beam would appear to be traveling along a curved trajectory. But again, according to the Principle of Equivalence, we can replace the accelerating elevator by a gravitational field. Viewing things this way, we would say that gravity bends light rays. Indeed, Einstein predicted that the Sun's gravitation would bend starlight by a tiny, but precisely determined, amount. This was first tested in 1919 during a total eclipse of the Sun. The result confirmed Einstein's prediction and made him a world celebrity. Furthermore, we define a straight line as the path of a light ray in a vacuum. No line is straighter·than this. If we could make a triangle out of three light rays, we could test whether this triangle was Euclidean—whether the sum of its interior angles was precisely 180 degrees. If this measurement were made in the presence of gravity, the light rays would not form a Euclidean triangle. Space would have become curved and its geometry non-Euclidean. Just imagine—all of this grew out of Einstein's thinking about an elevator.

The late I.I. Rabi once told me that he thought theoretical physicists were the "Peter Pans" of the human race. In a sense, they never really grow up. Newton was 24 when he contemplated the falling apple, and Einstein 28 when he contemplated the falling elevator. Both men still had a childlike capacity for wonder. A child sees deep and fascinating mysteries in things we adults find commonplace. That is one of the reasons children are so appealing. Scientists like Einstein and Newton never lose their sense of wonder. That is why learning about deep science is so satisfying. It puts us in touch with our childhood sense of wonder. It is something one wants never to lose.

Einstein's Scientific Legacy

Great art leaves a legacy that, if its creators could experience it, would probably astound them. Once made available, the work begins to lead a life of its own. The creator becomes almost a bystander. This is also true of great science, especially Einstein's science. He would be amazed, one imagines, to read about the vigorous debates now taking place over the foundations of quantum mechanics, a subject he struggled with for much of his life, or of the triumphs of modern cosmology, which has at its base the work he started some eight decades ago. Given his deep reservations about quantum mechanics, one supposes he would also be deeply skeptical of current attempts to incorporate general relativity into the quantum theory as part of a Theory of Everything. My purpose here is to convey a feeling for these developments—Einstein's scientific legacy. This survey will necessarily be very selective, since to do justice to the whole enterprise would require an entire book—probably several.

Let us begin with the special theory of relativity, which is now over ninety years old. Einstein created it in 1905, when he was 26. It has become so deeply ingrained in present-day physics that one sometimes finds it difficult to realize just how bizarre it seemed to Einstein's contemporaries. Here is a quotation from one of them, W.F. Magie, who was a professor of physics at Princeton and a president of the American Association for the Advancement of Science. In 1911, he gave a presidential address to the Association in which he said, referring to the theory of relativity, "I do not believe that there is any man now living who can assert with truth that he can conceive of time which is a function of velocity or is willing to go to the stake for the conviction that his 'now' is another man's future or still another man's past." Matters are even more complicated than poor Professor Magie imagined. Here are a few comments about what we have learned in the last few decades about the special theory of relativity—some of it very surprising.

One of the most important predictions of Einstein's special theory of relativity is what is called time dilation. If we have two identical clocks and set one of them moving with constant speed in a straight line, and if we use the stationary clock as the standard, we will find that, relative to the standard, the moving clock will be retarded. Time—Professor Magie *pace*—is a function of velocity. One way this was confirmed experimentally, nearly half a century ago, was by comparing the rates of decay of two identical particles—one moving, the other stationary. The rate of decay of a particle is a kind of clock. If the decaying particle has a lifetime τ as measured in a system at rest with respect to it, it will have a longer lifetime τ' as measured in a system moving with a speed v relative to the resting system—a lifetime expressed, according to the theory of relativity, in the relation

$$\tau' = \frac{\tau}{\sqrt{1 - v^2 / c^2}}$$

Here, c is the speed of light in a vacuum, which is 299,792,458 meters a second. It is the ratio of v to c that determines the magnitude of relativity effects.

This relation is tested successfully millions—indeed billions—of times daily in physics laboratories all over the world where decaying particles are observed. But these tests do not involve actually *looking* at a moving clock. In the case of the decaying particles, what one measures is the length of a track the particle leaves in a detector. If the particle lives longer than the classical theory would predict, it can move farther and its track is extended.

But what would happen if we actually *looked* at a moving clock through a telescope? What would we see? Looking at a clock means, for example, that photons emitted by its dial ultimately reach our eyes. A real clock is very complicated to analyze so let us replace it with an idealized situation. Suppose our "clock" is just a single atom that can emit a photon with a frequency v. That frequency is now our timer. Now suppose this "clock" is set into motion towards us. We know that the frequency of the light emitted by the atom will be Doppler shifted toward the blue. The frequency will be *higher* than the frequency of light emitted by the same atom at rest. Hence this "clock" will be seen to run faster! But then what is all this business about time dilation? Was Einstein wrong?

In order to understand what is going on we have to be clear about the difference between the arrangement Einstein imagined to measure time in the two systems, and the one we have just described, where we insisted on *looking* at the clock. We imagined a situation where photons from the clock were actually arriving in our eyes. In Einstein's arrangement we imagine putting, for the sake of illustration, identical clocks at all points along a railroad track. We also imagine that the conductor on a uniformly moving fast train has a clock of the same construction. As the train moves along, the conductor continually transmits to observers on the ground the time that he is measuring on his clock. These observers will find that the conductor's time is continually falling behind theirs by an amount determined by Einstein's formula. But *they* are not looking at the conductor's clock. *He* is looking at his clock, which is at rest in his hands. This is what Einstein meant by time dilation. There is no question of a Doppler shift here. There is no contradiction; the two situations are completely different.

While we are on the subject of time dilation, I want to note something about the Doppler shift in relativity. Here is a description of a very beautiful demonstration of the Doppler shift in sound that was apparently invented by Ernst Mach—at least he made a great deal of use of it. Imagine that one has a tube that can function as a whistle when air is blown through it. Imagine further that this tube can be made to turn around like an airplane propeller while it is whistling. Now, if you stand in the plane of the rotation you will hear the sound from the whistle being Doppler shifted as it spins. But what will happen if you stand on the axis of the rotation at right angles to the emitted sound waves? According to classical physics, you will hear *no* Doppler shift. This is also true, classical physics claims, if you replace this sound generator by a rotating light generator. (Sound and light must be treated quite differently in relativity. For sound, there is an "ether"—namely the medium in which the sound waves are vibrating. Unlike light, sound cannot propagate in a vacuum. The velocity of sound will be different for different uniformly moving observers.)

But relativity predicts something quite new for the Doppler shift in light. It predicts that even if one is at right angles to the motion of the source there will still be a Doppler shift of the light. This is a pure relativity time dilation effect. The first people to measure it were H.E. Ives and G.R. Stilwell, in 1938. It appears as if these experimenters—even in 1938—still did not believe relativity theory. They thought they were measuring something having to do with the ether! Nonetheless, despite themselves, they found the predicted relativity effect. By the 1960s, experimental techniques had improved to such a degree that experimenters were able to carry out a version of Mach's old experimental demonstration with light rather than sound. In these experiments the receiver was placed on the rim of a rapidly rotating disc, and the source at the center. From the point of view of the receiver, the motion is always at right angles to the incoming radiation. Again the novel relativity Doppler shift was observed. But the best demonstration of time dilation was carried out in the 1970s, when extremely accurate cesium clocks were taken aloft in airplanes and compared to identical clocks that stayed on the ground. In these experiments, several relativistic time-dilation effects are measured at once, including some from gravitation. All agree with the theory.

Einstein's theory also made a prediction about the lengths of uniformly moving objects. If we call the length of the object at rest L and the length of the same object in motion L', then the theory of relativity predicts the relation

$$L' = L\sqrt{(1 - v^2 / c^2)}.$$

For historical reasons this is called the Lorentz contraction—a reference to the Dutch physicist Hendrik Lorentz, who suggested it before relativity. Up until 1959, four years after Einstein's death and over 50 years after he created the special theory of relativity, if one had asked physicists the question "Can you *see* the Lorentz contraction?", I am sure almost all of them would have said "Yes." One wonders what Einstein would have said. The image that immediately comes to mind is that of a shrunken moving ruler or a distorted cube. But in 1959, a physicist named J. Terrell proved everyone wrong. The first reaction was disbelief, but now his result has entered into the textbooks—at least the good ones. As with time dilation, we must be clear what we mean by "seeing." It is a matter of insisting that the photons emitted by a moving object actually enter our eyes or are registered on a photographic plate.

At work here is a phenomenon that has been known for a very long time—"aberration." The standard illustration of aberration involves falling raindrops. To make things simple, suppose you find yourself standing in a rainstorm and all the raindrops are falling straight down from the sky. Now you decide to walk. As you do so, you will notice that the raindrops are no longer falling straight down but appear tilted at an angle. This is because you are observing the motion of the raindrops from a moving coordinate system and the angle of the velocity has changed. The knowledge that such an effect exists for light dates all the way back to the beginning of the 18th century. In 1727, a British astronomer named James Bradley noticed that as the Earth moved around the Sun, the apparent position of a star—he used α-Draconis—also moved. It traced out a tiny ellipse. He realized that this was because the angle of the incoming light changed during the course of the year due to aberration.

In classical physics, the angle of aberration is expressed simply by v/c, where v is the speed of the observer and, as usual, c is the speed of light. This formula applies when the angle of aberration is small. In Einstein's 1905 paper, he derived a relativistic expression for the angle of aberration using the Lorentz transformations. But no one realized—unless Einstein did secretly—what this implied about observing the Lorentz contraction itself.

To understand this, imagine a ruler placed on a line running parallel to some axis. Let us also suppose the ruler is luminous. We see the ruler because photons coming from its various parts reach our eyes. For these photons to reach us simultaneously, some will have left the ruler earlier than others to compensate for the different distances they have to travel; the ones from the back of the ruler will have left earlier than the ones from the front. Each of the incoming photons will travel on a straight line at slightly different angles; photons from the back of the ruler will travel on a trajectory that is at a smaller angle than photons from the front. Now, suppose the ruler is set into uniform motion along this axis. Due to aberration, all the photons will have their arrival angles shifted by small amounts that depend on the speed of the ruler. The effect is that the ruler will appear to have rotated. It will no longer appear to be parallel to the original axis. This result also holds for the classical theory of aberration.

But what Terrell found is that there is no other effect in relativity. All one will see is the rotation. There is no distortion of the kind that one might naively have thought would be produced by a Lorentz contraction. A moving cube, for example, will still look like a cube— but rotated. Amazing! More amazing is that no one thought of it for 50 years.

Now to the quantum theory, which Einstein also played a major part in creating. In the same year he invented relativity he published a paper on the quantum. He thought incessantly about it for the next two decades. After Max Born's probability interpretation was accepted at the end of the 1920s by most physicists, Einstein decided, as he wrote Born, that "it was not the true Jacob" and began to look elsewhere, although, as a kind of parting gift, he created the quantum theory of gases. But between 1926, when Einstein published his last paper on quantum gases, and 1935, when he dropped at least a momentary bombshell on the physics community, he published only one paper on quantum theory. It was published in 1931 in collaboration with R.C. Tolman, the distinguished cosmologist, and Boris Podolsky, one of the men with whom he would co-author the "bombshell" a few years later. What is noteworthy about the 1931 article was that it was Einstein's first technical paper written in English. It was published in the *Physical Review*, an American journal. After 1935, all of Einstein's technical papers were written in English and published in English-language journals. The 1935 "bombshell" was entitled "Can Quantum Mechanical Description of Physical Reality Be Considered Complete?". It was written with Podolsky and a third collaborator, Nathan Rosen. Forever after, it was known as the EPR paper. What was it about?

To understand this, it is useful to introduce a parable invented by John Bell. It is to Bell, more than anyone else, that we owe the current reawakening of interest in these questions. Until his death, Bell worked at the CERN laboratory in Geneva. He had a colleague there named Reinhold Bertlmann. After observing Bertlmann for some time, Bell noticed that he always wore socks of different colors. For the sake of discussion, let us suppose that Bertlmann owned only pink and green socks. Hence, if one observed Bertlmann wearing a green sock, one could be sure that the other was pink, even if one did not actually see it. In the somewhat more ponderous language of the EPR paper, we can say that the observation of

Bertlmann's green sock means that "there exists an element of physical reality corresponding to" Bertlmann's pink sock. We have no doubt that such a sock exists and we can readily imagine an observation that would reveal both socks simultaneously. Now comes the EPR argument.

Suppose we have two particles that approach each other from a great distance, interact, then separate and move off to a great distance. Like the colors of Bertlmann's socks, EPR imagined— these were thought experiments, although I will shortly tell you how they were ultimately made real—an experiment by which the measurement of the position of one of the particles implicitly measured the position of the other without disturbing it. Hence, they insisted, this would reveal the "physical reality" of the position of the distant particle. Then EPR invented a second experiment that similarly measured the *momenta* of another pair of particles. Hence, their combined experiments revealed, they insisted, the "physical reality" of the positions and momenta of the distant particles. The very fact that one can design these experiments, they would argue, means that position and momentum are at all times as real as Bertlmann's two socks. But this appears to violate Heisenberg's uncertainty principle, which states that position and momentum can never have a simultaneous "reality." To assign them a preexisting reality, like Bertlmann's other sock, is to go beyond the limitations of the theory. Hence, EPR concluded that quantum mechanics cannot offer a complete description of reality; there must be something else—some deeper reality.

When Niels Bohr, who as much as anyone guided the development of the theory, first saw this paper he was beside himself. Bohr was then in Copenhagen, collaborating with a physicist named Leon Rosenfeld. Rosenfeld later described what took place:

This onslaught came down upon us as a bolt from the blue. Its effect on Bohr was remarkable. . . as soon as Bohr had heard my report of Einstein's argument, everything else was abandoned: we had to clear up such a misunderstanding at once. We should reply by taking up the same example and showing the right way to speak about it. In great excitement, Bohr immediately started dictating to me the outline of such a reply. Very soon, however, he became hesitant: "No, this won't do, we must try all over again. . .we must make it quite clear. . . ." So it went on for a while, with growing wonder at the unexpected subtlety of the argument. Now and then he would turn to me: "What *can* they mean? Do *you* understand it?" There would follow some inconclusive exegesis. Clearly, we were further from the mark than we first thought. Eventually, he broke off with the familiar remark that he must sleep on it.

Rosenfeld goes on:

The next morning he at once took up the dictation again, and I was struck by the change in the tone of the sentences: there was no trace in them of the previous day's sharp expressions of dissent. As I pointed out to him he seemed to take a milder view of the case. "That is a sign," he said, "that we are beginning to understand the problem." And indeed, the real work now began in earnest; day after day, week after week, the whole argument was patiently scrutinized with the help of simpler and more transparent examples. Einstein's problem was reshaped and its solution reformulated with such precision and clarity that the weakness in the critics' reasoning became evident, and their whole argumentation, for all its false brilliance, fell to pieces. "They do it 'smartly,'" Bohr commented, "but what counts is to do it right."

Bohr was notorious for the length of time he took to write papers. He wrote, and he rewrote, and then he rewrote some more. Many of his papers might never have been written at all if Bohr had not had the services of brilliant young collaborators like Rosenfeld, who took down what Bohr had to say and served as sounding boards for his ideas. In this case Bohr was so agitated by the EPR paper that he finished his rebuttal in about six weeks—a sort of speed record for him. In fact, he was so concerned that he actually wrote two papers, a short precursor that was published in the British journal *Nature* and a longer one he wrote for the *Physical Review*. In the end, despite Bohr's comments, there is nothing really "wrong" with the EPR paper. If one could construct the apparatus needed to do their experiments, it would function as they said it would. But the point that needs to be emphasized—and this is basically what Bohr did—is that "experiments," plural, are involved. One could construct an EPR apparatus that would, implicitly, perfectly measure the momentum of a distant particle. But then this same measurement would, according to Heisenberg's uncertainty principle, give no information whatsoever about the particle's position. Then one could use a second piece of apparatus to, implicitly, measure perfectly the position of another distant particle. But this measurement would give no information whatsoever about its momentum.

This is an example of Bohr's notion of "complementarity." Different experiments can reveal different aspects of a quantum mechanical system, but a single experiment has limitations, imposed on it by the uncertainty principle, to what it can reveal. In the usual interpretation of quantum theory, the Copenhagen interpretation, the part of the system that the uncertainty principle says cannot be measured does not exist! It isn't that there is a well-defined momentum which the perfect position-measuring apparatus cannot get at. Rather, under these circumstances it is meaningless, according to the Copenhagen interpretation, to talk about momentum at all. In this sense the situation is quite different from Bertlmann's socks. With Bertlmann, we are persuaded that if we see a pink sock, the green sock exists and is on the other foot. If we design a suitable experiment—such as opening our eyes—both socks will be visible to us at the same time. Putting the matter perhaps too crudely, it is the difference between Bertlmann's socks and quantum mechanics that Einstein could not accept. He was convinced that at some deeper level there was a theory which gave as much legitimacy to the existence of the simultaneous position and momentum as common sense endows to Bertlmann's other sock.

While the EPR paper caused some stir at the time, most physicists quickly lost all interest in it. Quantum theory worked too well—better than any other physical theory ever invented. Furthermore, Einstein had no real alternative to propose. This lack of interest in Einstein's "dilemma" persisted until fairly recently. In the early 1980s I examined seventeen standard quantum mechanics books to see what they had to say about the EPR paper. Of the seventeen, only one—and I will come back to it shortly—made any reference to it at all. I am not talking about a lack of discussion. There is simply no reference to the paper at all. Even P.A.M. Dirac's great text *Principles of Quantum Mechanics,* with all its editions, has nothing to say. When asked about this, Dirac replied, in his slightly mysterious way, "I think it is a good book, except for the absence of several introductory chapters." These were the chapters where, one supposes, Dirac might have discussed such things as the EPR paper. At the time he first wrote the book, and when preparing subsequent editions, he apparently thought these matters were of insufficient importance to be discussed or even referred to.

The one exception is a book entitled *Quantum Theory*, published in 1951. It was written by the late David Bohm, who died in 1992. Connected with it is a remarkable tale involving Einstein. Bohm, who was born in 1917, developed as a young man a sympathy for Marxism, as did many bright young people of his generation. It is not clear that he was ever a member of the Communist Party, but it is quite clear—this was even established in a court of law—that he was never disloyal to his country. However, he was influenced by the Marxist notions, first formulated about modern science by Lenin and then adopted by other Soviet authorities, that modern science such as relativity and later quantum theory had dangerous elements of "idealism," which had to be watched with great care so as not to corrupt young Soviet scientists. (The situation was not as extreme in physics as it was in genetics. Until fairly recently, people who tried to teach genetics in the Soviet Union could find themselves imprisoned—or worse.) Physicists managed to continue using these theories by occasionally making a few, probably cynical, references purportedly showing their political correctness.

It appears Bohm was sufficiently influenced by Marxist theory that he was, at least before writing his book, somewhat uneasy about quantum mechanics. In his book, which is a really excellent introductory text, there is not a trace of this. In fact the book is one of the best defenses of the Copenhagen interpretation of quantum theory—the gospel according to Bohr—that has ever been written. Then came Einstein. It seems Einstein had read the book and then asked to see Bohm, who was at Princeton University. (In 1951, Bohm was forced to leave Princeton after he was indicted for contempt of Congress for pleading the Fifth Amendment at a hearing of the House Un-American Activities Committee in 1949. He was later acquitted, but could find no other job in the United States. Eventually he settled in

England, where he became a professor at Birbeck College of the University of London.) This was a happy coincidence because Bohm had also wanted to see Einstein to get his comments. One would give anything to have overheard the conversation. By the end of it, Bohm seems to have decided that the old man had a point. All his previous doubts about quantum theory came back, and Bohm set out to find a substitute that would include all the desirable features and none of the undesirable ones.

In his original book, which contains no trace of this discussion, Bohm not only explained the EPR experiment but invented a new version of it. It is vastly simpler than the original EPR setup—that must have appealed to Einstein—and it is the one on which all current work is based. It hinges on a property of particles like electrons known as "spin." Spin is a truly quantum-mechanical concept with no real classical analogue. But for the purposes of our discussion we can think of spin as being what it sounds like. It is an angular momentum that a particle can have even when it is at rest. But unlike a spinning top, electron spin can take on only two values, which we can call "up" and "down." These values can be measured if we put the spinning electron into a magnetic field. The direction that the electron, which itself acts like a tiny spinning magnet, moves under the influence of the field depends on the relative orientation of the field and the spin. Now we can imagine doing the following: We take two electrons and arrange things so that they are in a quantum-mechanical state which has a net spin of zero. The combined system then has no angular momentum. There are two ways this can happen; one electron can have spin up and the other down or vice versa. In the absence of prior knowledge—an experiment—we cannot tell which of these configurations applies. The quantum-mechanical description of this state contains both possibilities. Having prepared this state we now allow the electrons to separate. One stays here on Earth and the other, say, goes to the Moon.

In both places we have set up our spin-measuring magnets. To start, suppose we orient their fields in the same direction—call it "north" for want of anything better. Now suppose the Earth magnet indicates that our local electron has spin up. We can then guarantee that the Moon electron will have spin down. This is as sure a thing as Bertlmann's socks. If we are EPR people we can say that an element of "physical reality" has been conferred upon the spin down value of the Moon electron, which we never actually observe. Now we can rotate the two magnets so they point, say, west, and repeat the experiment with a different pair of electrons. Once again we will find that if the Earth-spin points in one direction, then the Moon-spin will point in the opposite direction. Again, if we are EPR people we will say that we have conferred an element of "physical reality" to the west component of the Moon-electron's spin. But here we have another Heisenberg uncertainty principle to contend with. It says that no two components of an angular momentum can be measured with arbitrary accuracy in a single experiment. By now we know the correct quantum-mechanical response: This is not one but *two* experiments with the magnets in different orientations done on two different pairs of electrons. Each experiment has revealed as much about the reality of the electron spin as quantum mechanics allows. The rest doesn't exist. There is no contradiction with the uncertainty principle and nothing to worry about.

But before we get too complacent, let us imagine that we now set both the Earth magnet and the Moon magnet to point north. If we do this, we will find the expected correlations between the spins of any pair of electrons we observe. Now, we can do something sneaky. While an electron is traveling to the Moon, we can, in collaboration with the Moon observer, change the orientation of both magnets so they now point, say, west. If we do this, the correlation between spin up and spin down will be as perfect as before, although the magnets are pointing in a completely different direction which was set when the electrons were moving towards their respective observers. This bothered Einstein deeply. He referred to it as *"spukhafte Fernwirkungen"*—spooky action at a distance.

Whatever attitude one wants to take towards these distant quantum-mechanical correlations, they are facts—facts that can be, and are, confirmed in laboratories. At one end of the attitude spectrum is the traditional Copenhagen interpretation. Its advocates would say that this theory predicts the correlations perfectly and no more should be asked or expected. (In other words: What more do you want?) At the other end are the New Age mystics who see these distant correlations as evidence for faster-than-light communication, with all sorts of wonderful transcendental possibilities. I think this simply represents a misunderstanding of the physics. But that is another discussion.

After his session with Einstein, Bohm came away with neither attitude. He set out to find a reformulation of the quantum theory in which statistics and probability were not fundamental but matters of convenience, just as they are in classical statistical mechanics. All processes in such a theory would be fully deterministic and could be followed in any detail one wished. Remarkably, Bohm did manage to find such a formulation, but at a cost. Most physicists do not believe the cost is worth the reward. Indeed, most physicists are not sure what the reward is. Bohm's reformulation involves what physicists call "nonlocality." The nonlocality involved here means that to fully describe what happens, say, here and now at the site of our Earth magnet, we must have information about what is going on at the same time everywhere else in the universe. It is not clear how such a requirement could ever be made compatible with the theory of relativity. Indeed, so far, Bohm's theory has not been generalized to include the requirements of relativity. In any event, Einstein was not very enthusiastic about the theory. Shortly after it was published in 1952, he wrote to his old friend Max Born "Have you noticed that Bohm believes (as de Broglie* did, by the way, 25 years ago) that he is able to interpret the quantum theory in deterministic terms? That way seems too cheap to me. But you, of course, can judge this better than I."

When Bohm's papers came out in 1952, John Bell was working in a substation for the British Atomic Energy Research Establishment in Malvern in Worcestershire. He was about to get his Ph.D. and was much taken by Bohm's ideas. He had become quite discontented with the sort of standard presentations of quantum theory one finds in Dirac's book. But it was only in 1963, while at Stanford University on leave from CERN, where he had moved permanently in 1960, that he found the time to really focus on these questions. This was not a very fashionable subject for a mainstream physicist to be working on at the time. Bell's work eventually made it fashionable. The problem that troubled Bell: Was Bohm's nonlocality essential to any attempt at replacing quantum mechanics by a deterministic theory, or was it an artifact of Bohm's specific models. As Bell once put it,

> [Bohm's] theory was nonlocal. Terrible things happened in the Bohm theory. For example, the trajectories that were assigned to the elementary particles were instantaneously changed when anyone moved a magnet anywhere in the universe. I decided to find out if this was a defect of his particular picture or was somehow intrinsic to the whole situation.

A person who had never been exposed to quantum theory might be tempted to say, upon first encountering the EPR spin correlations,

*Louis de Broglie, one of the early creators of quantum theory

that something must be missing in our description of the electron. The electron must have some hidden properties that determine these correlations. Bohm's theory contains such "hidden variables"—the term of art. It is these variables that obey Bohm's grossly nonlocal equations. Bell's question then became, is this a defect of Bohm's particular hidden-variable model or is it a common feature of all hidden-variable theories that reproduce the results of quantum mechanics? If it isn't a feature of the general situation, then, Bell reasoned, it might be possible to find a theory that would come closer to fulfilling Einstein's hopes. Bell chose to analyze the EPR experiment in the Bohm version with the spinning electrons since it was relatively simple and explicit.

Bell discovered that he had no problem producing a local theory of hidden variables which reproduced the quantum-mechanical result in situations where the magnets are pointing in the same direction—north, say—or when they are pointing at right angles to each other—east and north, say. In this case there is no correlation between the spins. But then he did something that no one else seems to have thought of. He allowed the angle between the two magnets to be anything, not just zero or 90 degrees. When he did this, he found something remarkable. It is now known as "Bell's theorem." Quantum mechanics makes a definite prediction for how the electron spins will correlate if the measuring magnets are at some arbitrary angle to each other. The prediction that a theory makes with local hidden variables is different for each theory. But, and this is Bell's theorem, *any* such theory would make a prediction that is *different* from quantum mechanics. In other words, one cannot have both the results of quantum mechanics and a *local* hidden variable theory; one or the other, but not both.

Bell published his result in 1964 in the first issue of an extremely obscure journal called *Physics,* which went out of business after a few issues. For five years no one seemed to pay any attention to it and Bell went back to his regular work as an elementary particle theorist and an accelerator designer. However, in 1969, four physicists, J.F. Clauser, M.A. Horne, A. Shimony, and R.A. Holt, published a paper in which they proposed a real way to carry out the hypothetical EPR experiment and, using Bell's theorem, to see if quantum mechanics was right. They also found a generalization of Bell's original theorem that was easier to test. Furthermore, instead of correlated electrons they proposed to use correlated photons—particles of light.

The property of the photon that replaces electron spin is known as "polarization." When a light wave propagates through space the direction of the wave oscillation is always at right angles to the direction of its propagation. If the wave is traveling north, the oscillations may be either to the east or the west or they may rotate to the right or the left, always at right angles to the direction of motion. Such a wave is called "transverse," and the two situations we just described are referred to as "plane" or "circular" polarizations, respectively. If two photons are produced in a suitable set of atomic transitions, their polarizations can have exactly the same kind of correlations as the electron spins in Bohm's version of the EPR experiment. But photon polarizations are measurable not only in principle but in practice. Experiments to measure these correlations began in 1972 and continue today in laboratories all over the world.

This is a story that is still being written, but at least until now, all the evidence is consistent with the correctness of quantum theory. Local hidden variables appear to be ruled out. However, almost every issue of our physics journals now contains one or more paper about the EPR experiment and the generalizations and subtleties of Bell's theorem. Not only that, but the whole subject of the meaning and interpretation of quantum theory has now become "respectable." Very distinguished physicists, who might have been afraid of being called "senile" for worrying about such matters, now write papers on the foundations of the theory. This is part of Einstein's legacy. One can imagine he would find it very amusing that after nearly 60 years, his old paper has been given a new life.

Next we turn to general relativity and cosmology. In 1919, Einstein's prediction, from his then-new theory of gravitation, that starlight would be bent by the gravitational field of the Sun was confirmed by two British solar eclipse expeditions. His prediction of a tiny bending angle of 1.74 seconds of arc is in essence what these observers found. Below is a partial list of more recent eclipse measurements which will give the flavor of what has been happening in this study since 1919. The first entry refers to the sponsoring observatory; the second, the location of the eclipse; and the third, the result of the observation followed by the claimed experimental error. The results and the errors are in seconds of arc.

OBSERVATORY	LOCATION	DATE	RESULTS ± ERROR
Greenwich	Australia	Sept. 21, 1922	1.77 ± 0.40
Potsdam	Sumatra	May 9, 1929	2.24 ± 0.10
Sternberg	U.S.S.R.	June 19, 1936	2.73 ± 0.31
Yerkes	Brazil	May 20, 1947	2.01 ± 0.27
Yerkes	Sudan	Feb. 25, 1952	1.70 ± 0.10
Texas	Mauritania	June 30, 1973	$(0.95 \pm 0.11) \times L_E$

(In the last entry, the result is given as a percentage of the Einstein prediction L_E.)

It is clear that these values have a good deal of scatter, which shows just how hard these observations are. But overall there is good agreement, within the experimental errors, with Einstein's prediction.

Einstein's theory differs dramatically from Newton's. In Newton's theory one would say that the sunlight was bent by the force of gravity acting on it. In Einstein's theory there is no force of gravity in the usual sense. What happens is that gravity "curves" space—and changes time. The light moves in a geometry determined by the gravitation. This is what produces the bending. But bending can be seen in even more direct ways. These involve what are known as "quasars," an abbreviation for "quasi-stellar source." A quasar is a source of energy at the edge of the visible universe, which must represent a state of matter as it was billions of years ago.

In 1968, a special quasar, which like the others is a light year or two in size but 1,000 times more luminous than a giant galaxy, was found to have a delightful property. This quasar, designated 3C279 in the Cambridge University catalogue of such things, is eclipsed by the Sun each October. This is a kind of private eclipse that can be studied closer to home on an annual basis—unlike the traditional solar eclipses that require distant expeditions to get to them. Quasar 3C279 emits radio waves, which are bent by the Sun just as light rays are. There is also a neighboring quasar known as 2C273. Its apparent position relative to 3C279 can be measured just before and after the eclipse. The apparent shift in position due to gravity—the bending of the radio waves—again agrees with general relativity.

There is an even more precise test for general relativity effects using radar. In the 1960s I.I. Shapiro, then at M.I.T., realized that if a radar signal was sent to a planet like Mercury when Mercury was in a position in its orbit which required that the signal pass close to the Sun, then according to general relativity, the signal would be delayed by the Sun's gravity, compared with the time required for it to travel to the planet if the Sun was not there. The time delay is so small that in order to measure it by bouncing the signal off the planet and then timing the round trip, one has to know the distance from here to Mercury to an accuracy of less than two kilometers in some one hundred million! The general relativity delays are only on the order of 200 microseconds. But radar techniques are so accurate that these tiny delays have been observed. They also agree with the general relativity prediction.

In 1936, Einstein made another general relativity prediction, which for practical reasons he thought would be of only academic interest. But it too has become a very active area of research. It involves what is known as "gravitational lensing." He published a brief note in the American journal *Science* entitled "Lens-like Action of a Star by Deviation of Light in the Gravitational Field." In this note Einstein imagined that, by some miracle, two stars bright enough to be visible to us would somehow line themselves up perfectly, one behind the other. In the absence of the gravitational effect that Einstein was calling attention to, an Earth observer would simply see the front star, which would appear as a point of light. The back star would be hidden. But general relativity teaches us that gravity bends light rays. Hence, the light from the rear star will be bent around the front star on all sides. But this is what a lens does. A lens is made of some glass-like material that bends light as it enters it from the air. If we look at the light through the lens, the source of the light will look distorted—magnified, for example. Einstein pointed out that if we could observe such a pair of aligned stars, then because of the gravitational lensing, we would see a ring of light produced by the rear star while the front star would appear as a point of light in the center of the ring. However, Einstein believed that this effect would never actually be observed because it was so small, and furthermore, two stars are very unlikely to line themselves up like this.

In 1937, astronomer Fritz Zwicky generalized Einstein's idea to aligned galaxies—fields of stars—where the effect should be much bigger. But it was not observed until 1979 (coincidentally the hundredth anniversary of Einstein's birth) and not quite in the form Zwicky had imagined. It was in the early 1960s that quasars were discovered. But in 1979, the British astronomer Dennis Walsh discovered a very peculiar pair of quasars. They were identical—twins—and seemed to be exactly the same distance from the Earth. It occurred to him that this might be a gravitational lensing effect. One might be seeing a single quasar, but one which was behind a galaxy that was acting as a lens. In fact, he then went on to identify the galaxy that was doing the lensing. These do not look like Einstein's rings. The sorts of visual effects the lensing galaxy will produce depend on its alignment with the quasar as well as the size and shape of the "lens." There can be the original Einstein ring effect, pairs of crescents, or even the apparent brightening of the quasar seen through the lens. All of these things have by now been observed, and gravitational lensing has become a small industry among astronomers.

The last aspect of Einstein's legacy that I want to discuss here is cosmology, the origin and destiny of the universe at large. I will reserve the subject of black holes for a separate essay, because Einstein's relation to black holes is so peculiar.

Einstein began applying the results of his theory of general relativity and gravitation to the study of the large-scale evolution of the universe very soon after he published it. His first paper on this subject was published in 1917. At the time, it was generally accepted that the Milky Way galaxy was, in fact, the entire universe, something we now know to be absurd. The problem Einstein was concerned with was what keeps this structure from collapsing due to the mutual gravitation of its component parts. He decided that nothing did, unless he modified his theory to add a compensating repulsive force of mysterious origin. That was the subject of his 1917 paper. However, a few years later a Russian polymath named Aleksandr Friedmann showed that Einstein's original theory contained expanding, contracting and even stationary universes. At first, Einstein did not believe this result. He even published a brief, incorrect paper which claimed to refute it. But he took it back and with considerable reluctance agreed that Friedmann's expanding universe was at least a mathematical possibility.

By 1929, however, the American astronomer Edwin Hubble, who had also found evidence that the Milky Way galaxy was only one of many, discovered that the galaxies were in fact expanding away from each other just as Friedmann had said they might. This persuaded Einstein to drop his old cosmological theory. [In the second (1945) edition of his wonderful lectures entitled *The Meaning of Relativity*, first delivered in Princeton in 1921, Einstein devoted an appendix to a discussion of Friedmann's equations. This seems to have been his last written comment on cosmology.] In 1927, the Belgian astronomer Abbé Georges Lemaître independently discovered Friedmann's equations, but in the mathematical form we currently use. He also raised the obvious question: If the universe is now expanding and we run the expansion backwards, how did it start? What caused the expansion to begin? Lemaître had the fanciful idea that, originally, the universe was contained in a "cosmic egg" which exploded—a Big Bang—producing the expansion we now observe.

This was not very quantitative. The real father of the modern quantitative Big Bang theory was the late George Gamow. Gamow died in 1968 so he did not get to see all of his ideas in full fruition. But he lived long enough to see some of them. Gamow was a wonderful Russian eccentric who escaped from the Soviet Union in the late 1920s. Ironically, shortly before leaving the country, he had enrolled in a course in cosmology that Friedmann was supposed to teach in St. Petersburg the year he died. After his escape, Gamow went to Bohr's institute in Copenhagen, where he immersed himself in the developing quantum theory and later in nuclear physics. In 1934, he emigrated to the United States and took a position at George Washington University, in Washington, D.C. After the war, in collaboration with some younger colleagues, he began to work out in detail what happened after the Big Bang.

Gamow's original idea was that all observable matter had a cosmological origin. In the beginning, Gamow claimed, there was a cosmic soup that he called the "Ylem"—from the Greek word *hyle* meaning either "wood" or "matter." Gamow's idea was to make the Ylem as primitive as possible and to see if all the matter we know about could have evolved from it. His Ylem consisted of particles like the electron, the photon and the elusive neutrino, the massless, chargeless particle that barely interacts with anything. There were also protons and neutrons in equal numbers and their antiparticles as well. In the modern cosmologies, quarks have replaced the protons and neutrons in the Ylem. (In this version, part of what has to be explained is how the quarks, which are the fundamental constituents of particles like protons and neutrons, came to form them.) To a modern cosmologist, Gamow's Ylem with its neutrons and protons, is appropriate to the early universe but not to the *very* early universe. Nonetheless, the questions that Gamow and his young colleagues struggled with are, in many instances, the same ones we struggle with now.

Gamow wanted to show how the particles in his Ylem could combine into the elements we now see. The basic mechanism for this combining process is nuclear fusion. Fusion occurs when two light elements combine into other light elements that are more tightly bound than the fusing elements. To take the simplest example, a neutron and a proton can combine to make a deuteron, the nucleus of "heavy" hydrogen. The deuteron is less massive than the neutron and proton together, so that the excess mass-energy is released in the form of a gamma ray—a very energetic quantum. The loss of mass is converted into energy according to Einstein's equation $E = mc^2$. In the Sun, a series of fusion reactions combines four protons into a helium nucleus with the release of the solar energy we live by. Gamow wanted to build up all the elements this way but he was stopped by nuclear physics, more or less at helium. The heavy elements, we now believe, are created in supernova explosions and do not have a cosmological origin. On the other hand, Gamow and his associates were able to explain why the universe contains so much helium. About a quarter of the nuclear mass in the universe is helium and it surely has a cosmological origin.

Along with his young associates R. Alpher and R. Herman, Gamow also investigated what had happened to the radiation produced in the Big Bang. They reasoned that the early universe acted like a giant cavity which confined the radiation. Any hydrogen atom that formed was likely to be torn apart by a photon in the background radiation. After about 100,000 years, the universe cooled off enough for electrons and protons to combine and form hydrogen atoms. At this point, the Big Bang photons can no longer interact with matter so they continue to cool as the universe expands at a rate given by Friedmann's equations. Today, Gamow and his colleagues reasoned, we should observe a background of cosmic radiation at a temperature of a few degrees above absolute zero. This prediction was made in the 1940s and then forgotten or ignored. In 1965, A. Penzias and R. Wilson from Bell Telephone Laboratories discovered the background radiation accidentally when it appeared as "static" in a radio telescope they were using. Thanks to recent satellite measurements, the radiation temperature is now very accurately known; it is 2.726 ± 0.002 degrees above absolute zero with the small error shown in the last decimal place. This means that there are, on average, about 417 cosmic photons in every cubic centimeter of the universe. This discovery initiated the modern, and currently very active, science of quantitative cosmology. Although Einstein did not live to see any of it, this too is part of his legacy.

Everywhere we turn in modern physics, from the laser to the black hole, we see the hand of Einstein. For a 20-year period, from 1905 to 1925, he seems to have had a direct pipeline to the "Old One"—the guardian of the secrets of nature. During this period, Einstein's physical intuition was so powerful that almost everything he touched in physics turned to gold. This was, and he said it often, a source of great joy to him. But his success and the adulation it brought also seem to have puzzled him. One gets a sense of this from a fragment of a conversation that Robert Oppenheimer recalled having with him on Einstein's 71st birthday. As Oppenheimer walked Einstein back home from the Institute for Advanced Study that day, Einstein said, "You know, when it's once been given to a man to do something sensible, afterward life is a little strange."

Bohr

Niels Bohr had the characteristic, rare among both scientists and people in general, that anyone who came into contact with him, even very briefly, was deeply impressed. Not all these impressions were favorable. Winston Churchill, who met Bohr after the war—an occasion during which Bohr tried to advance his ideas on nuclear disarmament—thought he was a dangerous subversive who probably should be locked up. My own very brief encounters with Bohr took place in 1958, when I was at the Institute for Advanced Study in Princeton. One of them involved giving a brief presentation, in a seminar, of some work I had just done with T.D. Lee and C.N. Yang. My two very senior, soon to be Nobelist, collaborators had their own work to present to Bohr, so I was left to carry our joint spear. I was so nervous that I went through my allotted ten minutes in about one. If Bohr understood what I was talking about, he did not let on, nor did he ask any questions. Later, I wondered if he had connected me to an incident that had occurred a few months earlier. Marvin L. Goldberger— "Murph"—and his wife, Mildred, had given a New Year's Eve party. Goldberger was then a professor of physics at Princeton University. Subsequently, he became the president of Cal Tech and later the director of the Institute for Advanced Study. In any event, for this occasion, he and Mildred had rented a local firehouse. At the time, I was still playing the trumpet and Goldberger had requested that I play "Auld Lang Syne" at midnight. Just as I began my solo, in walked the J. Robert Oppenheimers and the Bohrs. Both Oppenheimer and Bohr gave me extremely fishy looks. Did Bohr connect the trumpet solo with my seminar?

Later that same year I had the opportunity to witness Bohr on an occasion I have never forgotten and which may carry some lessons for evaluating radical new ideas in physics. These were the pre-quark days, and new, unanticipated elementary particles were appearing in a bewildering profusion in both cosmic ray- and high-energy accelerator experiments. Things had become so desperate that Oppenheimer (I *think* he was kidding) suggested that a Nobel prize be given to any experimental physicist who did *not* discover a new particle. In the confusion, a rumor reached the Institute that Werner Heisenberg and Wolfgang Pauli had discovered a Theory for Everything—each generation has one. By this time, Heisenberg's physics was thought to be a bit over the hill, but Pauli, who died later that year, was still considered one of the most brilliant and skeptical physicists who had ever lived. When Pauli said, as he occasionally did, that a paper was not *even* wrong, it generally sank without a trace. That Pauli was taking part in such an enterprise lent it *gravitas*.

Pauli was invited to lecture on this work at Columbia University in late January 1958. A group of us traveled from Princeton to hear him. I sat next to Freeman Dyson during the lecture. Not long after it began, Dyson said to me, "It is like watching the death of a noble animal." He had seen at once that the new theory was hopeless. What none of us knew was that Pauli would die of cancer a few months later. But before his death, he too turned against the theory and circulated a cartoon he devised which showed only a blank canvas and a caption, in Heisenberg's voice, which read "I can paint like Titian— only a few details are missing."

Bohr was also at the lecture. He was a big man, dressed in an elegant dark suit with a vest, who reminded me of a Saint Bernard. He was then in his early 70s. After Pauli finished, Bohr was called upon to comment. Pauli remarked that at first sight the theory might look "somewhat crazy." Bohr replied that the problem was that it was "not crazy enough." This was a very deep remark, disguised, as was often the case with Bohr, as a little joke. What he had in mind was that a really radical and profound theory, like quantum mechanics, appears at first sight to be totally mad. The idea that there are limitations to measurement of the kind expressed by the Heisenberg uncertainty principles looked to many people at the time—and even now—as being perfectly crazy. Heisenberg's new theory, on the other hand, seemed merely complicated. It was also, as it turned out, manifestly wrong. But then Pauli and Bohr began stalking each other around the large demonstration table in the front of the lecture hall. When Pauli appeared in front of the table, he would tell the audience that the theory *was* sufficiently crazy. When it was Bohr's turn to stand in front of the table, he would say that it wasn't. It was an uncanny encounter of two giants of modern physics. I kept wondering what in the world a non-physicist visitor would have made of it.

In the beginning of his biography of Bohr, *Niels Bohr's Times, In Physics, Philosophy and Polity*, the physicist turned historian of science Abraham Pais discusses the fascinating question of what is the lasting significance of Bohr's work in physics. He describes a conversation he had with someone he identifies only as "one of the best and best-known physicists of my own generation." This anonymous scientist says to Pais "You knew Bohr well," to which Pais replies "I did." "Then tell me," Pais's interlocutor goes on, "What did Bohr really do?" This is, in its way, an amazing question. The same physicist could not possibly have asked the same question about Einstein. If he had, the answer would be simple: Einstein created 20th-century physics. His papers were so deep and so complete than even now, in some cases almost a century after they were written, one could use them as a textbook to teach their subject matter.

I cannot, on the other hand, imagine anyone but a historian of science reading Bohr's original papers. Not only are they densely written—the act of composition caused Bohr an anguish that almost bordered on physical pain—but even those that are much more recent than Einstein's seem dated. The philosophical papers, of which Bohr seems to have been especially proud, are all but unreadable. I once asked the late I.I. Rabi what he thought of Bohr's philosophical contributions to quantum mechanics. He said,

> This work was his life. There was no point in trying to tell him that I thought it was irrelevant to the sort of things that an experimenter actually does in the laboratory. I felt that he was very profound about things that didn't really matter. But one was not going to tell him that.

Nonetheless, there is something ironic and peculiar about all this. Einstein was, in the deepest sense, a "classical" physicist. His sensibilities were formed during the 19th century and he was never able to accept the quantum-mechanical view of reality. Bohr, on the other hand, was the guiding hand in creating this reality—the sounding board for Heisenberg, Pauli, Dirac, Schrödinger, and the rest of the inventors of quantum mechanics. Yet with rare exceptions, when one reads Bohr on quantum mechanics, he seems almost wholly obscure compared to Einstein. No one put this more clearly than the late John Bell, who among all the present generation of physicists, thought the deepest about quantum theory. Bell once said to me,

> I feel that Einstein's intellectual superiority over Bohr in this instance [the quantum theory of measurement] was enormous: a vast gulf between the man who saw clearly what was needed and the obscurantist. So for me, it is a pity that Einstein's idea [of classical, causal reality] doesn't work. The reasonable thing just doesn't work.

This having been said, why is Bohr in many people's reckoning, the most important physicist of this century after Einstein? It is much more difficult to explain this to a non-physicist than to explain why Einstein was the greatest physicist of the century, indeed the greatest since Newton. Einstein was dealing with fundamental issues of space and time, while much of Bohr's work was highly technical, and the parts dealing with quantum mechanics were immensely subtle. Not only that, some of what Bohr published turned out to be totally wrong. While one may argue that some of Einstein's very last work came close to the "not even wrong" category, for nearly 30 years everything the man published was a mine of gold, whose riches are still far from exhausted.

Bohr's seminal work was done in 1913, when he was 28. He had just returned to Denmark after a postdoctoral appointment in Manchester. His teacher in Manchester was the great New Zealand–born experimental physicist Ernest Rutherford. Rutherford and his young collaborators had, a few years earlier, discovered the atomic nucleus. Two of them, Hans Geiger (he of the counter) and Ernest Marsden, had been set the task of scattering so-called alpha-particles, actually helium nuclei, produced when radioactive decays of some heavy elements take place. They are then scattered by thin foils of gold. Much to the astonishment of the experimenters, some of the alpha-particles were scattered backward, as if they had struck something hard within the gold atom. They had. They had collided with the gold atom's nucleus. Before this, the most common view of the atom was that it was a fuzzy ball of electric charges. According to that picture, the alpha-particles were predicted to pass straight through the gold foil like bullets through fog. Instead, a few of them had bounced backward.

It took some 25 years before it was established that the nucleus consisted of relatively massive neutrons and protons, with the lighter particles—electrons—circulating outside. But almost from the beginning, it was realized that the new model of the atom with a nucleus was in serious conflict with classical physics. An accelerated electron—or indeed any electrically charged particle—radiates and hence loses energy. How then could matter be stable? Why didn't the atoms simply collapse as the electrons radiated away their energy? Moreover, according to the classical picture, the radiation from the electrons would be entirely chaotic. But in fact it was emitted in beautiful spectral lines whose wavelengths, in the simplest cases, were related to each other by elegant mathematical formulas, which had been discovered empirically.

Bohr solved these two problems with one masterstroke. In current language, he "quantized" the electron orbits. This means he advanced the hypothesis that the electron, as it moved around the nucleus, could move only in special orbits. We now refer to them as Bohr orbits. The orbit with the least energy, the so-called ground state, was assumed to be absolutely stable. When an electron transited from one orbit to another—made a quantum jump—quanta of radiation were given off, with an energy determined by the energy difference between the orbits. Since the allowed orbits follow discrete patterns and have discrete energies, so does the emitted radiation. Hence the spectral regularities.

All of this might have been dismissed as so much speculation if Bohr had not been able to make it quantitative. (In this I am reminded of Kepler, who was saved from the worst sort of Pythagorean mysticism by his determination to produce a quantitatively accurate description of the orbit of Mars.) Using his quantization, Bohr was able to compute the magnitude of the frequencies emitted, as well as their interrelations. When a tiny discrepancy appeared between Bohr's predicted values and the measured values, he was also able to account for it as an effect of the relatively slow motion of the ponderous nucleus. When Einstein heard of these results he remarked "This is an enormous achievement. The theory of Bohr must be right."

Bohr's atom, with its classical orbits, quantized in space, has become one of the defining images of the atomic age. We see drawings of it everywhere. Yet it does not correspond at all to the modern quantum-mechanical understanding of the atom. Rather, it is part of what is now known as the "old" quantum theory—an uneasy marriage between classical pictures of the atom and the quantum constraints. From the time Bohr published his papers until the mid-1920s, when the "new" quantum theory was invented, this ungainly structure was elaborated ad infinitum. One is reminded of Ptolemy's attempt to save the geocentric cosmology by adding more and more epicyclical planetary orbits, all of which were replaced when Kepler introduced elliptical planetary orbits around the sun.

Volumes have been written about the invention of the new quantum theory. From everything I have read, I am persuaded that it was Heisenberg who had the first truly quantum-mechanical mind. He recognized that Bohr's orbits were, in a certain sense, irrelevant. One cannot observe an electron making an atomic orbit. The act of observation destroys the orbit—"we murder to dissect"—since the electron is knocked out of the atom. This realization was later canonized by Heisenberg in his uncertainty principles. What one *can* observe are the spectral lines. Heisenberg concentrated on these, making a kind of calculus which became known as matrix mechanics. It was a Faustian bargain, since to accept this calculus was to give up visualizing the orbits. About the same time Schrödinger invented what appeared to be a second quantum theory in which the electron is described as a packet of waves. At first, people like Einstein who wanted to cling to classical realism were very pleased with this version of the theory since it appeared as if the waves could be visualized. This hope faded when the Heisenberg and Schrödinger theories were shown to be mathematically equivalent. Max Born then demonstrated that the Schrödinger wave packets had to be interpreted as packets of probability. The Bohr orbits were regions of space where the presence of the electron is highly probable.

Niels Bohr's role in all of this was, in a certain sense, Socratic. By the mid-1920s he had been given his own institute for theoretical physics in Copenhagen—it was officially opened in 1921. So he was in the position of inviting the new breed of quantum theorists to Copenhagen to work out their ideas. In these sessions Bohr was relentless. On one occasion, the exhausted Schrödinger was forced to take refuge in his bedroom, only to be followed there by Bohr, still arguing. Out of this came what is known as the Copenhagen interpretation of quantum theory, which curiously does not seem to have been written down systematically anywhere. Its two key elements were "correspondence" and "complementarity." Correspondence refers to the requirement that quantum theory have a classical limit in the sense that the experiments involving quantum mechanical processes can ultimately be described by using classical measuring instruments. For example, while the origin of the atomic spectra is quantum jumps, the spectra themselves can be photographed by an ordinary camera. Once again Bell put the issue succinctly:

> I disagree with a lot of what Bohr said. But I think he said some very important things which are absolutely right and essential. One of the vital things that he always insisted on is that the apparatus [the instrument used to measure wave and particle phenomena] is classical. For him there was no way of changing that. There must be things we can speak of in a classical way. For him it was inconceivable that you could extend the quantum formalism to include the apparatus.

One can always attempt to describe an individual measuring instrument using the quantum mechanics of its atoms and molecules, but then this description must refer to another classical apparatus on the next level in the hierarchy. At the end of the day, the language we use must be classical. It is the only language we know.

Complementarity, which became for Bohr a general philosophical principle, was conceived in response to the particle-wave duality of matter. An object such as an electron can exhibit both particle and wave characteristics, depending on the experimental arrangement used to measure its properties. On their face, these properties seem to exclude one another. Waves can pass through each other, modifying the patterns, while particles, as usually understood, cannot. Bohr noted that these properties are not really contradictory but rather complementary, since they can never be realized conjointly in a single apparatus.

Each experimental setup reveals a distinctive feature of the electron. The Heisenberg uncertainty principles, it turns out, prevent contradictory facets of physical phenomena from being revealed simultaneously. But Bohr was not satisfied with limiting the idea of complementarity to physics. He thought he saw it virtually everywhere; for example in instinct and reason, free will, love and justice, to name only a few of his applications. While some people profess to find these antimonies enlightening, I have always found that such applications of complementarity lead to a dead end.

I have also found the Copenhagen interpretation increasingly obscure, no doubt because of the influence of my friend John Bell. The question that appears to be unanswered within quantum mechanics itself is to what domain it is supposed to apply. It is all very well to talk about different kinds of "classical" apparatus, but what are they? How many atoms does it take to construct such an apparatus? As Bell put it:

> It is very strange in Bohr that, as far as I can see, you don't find any discussion of where the division between his classical apparatus and the quantum system occurs. Mostly you will find that there are parables about things like a walking stick—if you hold it closely it is part of you, and if you hold it loosely it is part of the outside world. He seems to have been extraordinarily insensitive to the fact that we have this beautiful mathematics, and we don't know which part of the world it should be applied to.

This is not to say that for all practical purposes—what Bell used to abbreviate as FAPP—this division cannot be made. For most physicists, certainly those of the older generation, this is quite enough. From the FAPP point of view quantum theory has successfully answered every question put to it. I am constantly reminded of a story Dyson told me many years ago concerning his young children. His daughter was explaining to her younger brother that she could row a boat because she understood how the "rowers" worked. To this her brother replied that he did not understand how the "rowers" worked, but he could row the boat anyway. There is a new generation of physicists who have been trying to find out how the "rowers" work. In my view, this enterprise is unlikely to lead very far until there is guidance from experiment. Every experiment that has been devised so far agrees with the predictions of quantum theory.

Bohr's early success in physics made him a celebrated figure in Denmark. He rapidly became a smiling public man. Unlike Einstein, he established a school and was very successful in raising the money to keep it running. One cannot imagine Einstein applying for grants or heading scientific associations. Nor can one imagine Einstein seeking out world leaders like Churchill and George Marshall to discuss his ideas about an open world in which knowledge about nuclear weapons would be freely exchanged among the United States, the Soviet Union and other countries. Bohr got a sympathetic hearing from Roosevelt, at least on their first meetings, and a very unsympathetic hearing from Churchill. Oppenheimer once thought of writing a play that he was going to call "The Day That Roosevelt Died." His point was that if Roosevelt had lived longer, some of Bohr's ideas about an open world might have gotten farther. In particular, the nuclear arms race, from which we now seem to be emerging, might have been averted. It is this last point I wish to turn to and finally to the question (recently raised) of whether Bohr actually tried to further his open-world agenda by giving nuclear secrets to the Russians.

In the last few years, with the end of the cold war and the opening of the Soviet archives, we have learned an enormous amount about the Soviet nuclear weapons program. But even before that, André Sakharov in his memoirs made the essential point: that nothing Bohr, Churchill, Roosevelt or anyone else could have said would have halted the Soviet program to first build an atomic bomb, then a hydrogen bomb. Sakharov wrote:

> The Soviet government (or more properly those in power: Stalin, Beria and company) already understood the potential for the new weapon and nothing could have dissuaded them from going forward with its development. Any US move toward abandoning or suspending work on a thermonuclear weapon would have been perceived either as a cunning, deceitful maneuver or as evidence of stupidity or weakness.

Sakharov might have added that the key Soviet physicists, such as Iuli Khariton, Iakov Zel'dovich, Igor Kurchatov and others, including Sakharov himself, needed no special threats or encouragement to work on nuclear weapons. They were genuinely terrified of American nuclear hegemony. They had just been through a war that came close to destroying their country and which they might well have lost. While they may not have shared Stalin's territorial ambitions, they did share his determination to break the western monopoly on nuclear weapons. The fact that Stalin held the pistol of his secret police chief, Lavrentii Beria, to their heads was more an irritant than a goad.

It should be added that Bohr's goal of sharing nuclear secrets with the Soviet Union had already been achieved, although not in the way he intended. What has been revealed in the last few years is that up until the time we began constructing the hydrogen bomb in earnest, there was almost nothing of real importance in our program that the Soviets had not learned through espionage. By the time Klaus Fuchs left Los Alamos, he had given the Soviets the complete blueprint of how to make a plutonium bomb—with the engineering details, which are what matter. This probably saved the Russians a few years in the development of their first atomic bomb, but they would have produced one anyway. The hydrogen bomb they did on their own.

The question is whether Bohr played any part in this espionage. The universal conclusion of people who have examined all the information objectively is "no." But the grounds for this suspicion are interesting. In November 1945, on Beria's instructions, a young Soviet physicist named Iakov Terletskii was sent to Copenhagen to learn whatever he could about the Allied nuclear weapons program from Bohr. Beria was concerned with the possibility that some of the information Fuchs was feeding his scientists might have been planted—it wasn't—and therefore Bohr, even if he didn't reveal anything new, might provide valuable confirmation.

Bohr received a request to meet with Terletskii from a communist member of the Danish parliament, and on November 14th the meeting was arranged. Whether Terletskii knew it or not, it was a meeting in a fishbowl. Prior to it, the British Embassy had been informed and they in turn informed General Groves in Washington. The Danes and Bohr's son Ernest provided armed protection in case the Soviets had any ideas about kidnapping Bohr and the entire meeting was witnessed by Aage Bohr, one of Bohr's other sons. Terletskii, who did not seem to know a great deal about the subject, had a list of 22 questions, prepared by the Soviet nuclear weapons scientists, to put to Bohr. A transcript of these, and Bohr's answers, has recently become available. It is a very odd document. The answers Bohr gave are, at all the crucial places, either evasive or wrong. One can interpret this in one of two ways. Either Bohr was playing a very clever cat-and-mouse game, or he really didn't know that much about the subject. I think both are probably true.

Before Bohr came to Los Alamos at the end of 1943 he knew nothing about the Allied program. He was briefed on some of the essential physics by Richard Feynman, who seems to have been assigned that task. There is no evidence that he knew anything about the engineering details, and he never visited any of the plants where the production of plutonium or the separation of uranium isotopes was taking place. His real role at Los Alamos was to attempt to set the nuclear program into a larger political and moral context. Most of the scientists working on the project had had no time to think about what would happen next—if the bomb worked. This is what occupied Bohr's time.

At a second meeting with Terletskii, Bohr handed him a copy of the so-called Smyth Report on the Allied nuclear weapons program. By this time, 100,000 copies had been sold and the Soviets had nearly finished their translation into Russian. I have reread this report carefully and once again have been struck by the fact that it tells almost nothing about how to make a nuclear weapon. Its primary aim was to show just how monumental a project it was. There is something ironic in this too. The Soviets built theirs because we had built ours. We built ours because we were afraid the Germans were building theirs. But the Germans, after trying, had come to the conclusion that they couldn't build one, and that therefore no one else could build one either.

The Drawing

In September 1943, Niels Bohr learned that the gestapo in Copenhagen intended to arrest him. A few weeks later, on the 29th, he, his wife, and several others hoping to escape from Denmark walked from the Bohrs' home in Carlsberg to a meeting place near the water. From there they crawled in complete darkness to a beach. They boarded a boat and secretly crossed the Oresund to Sweden. On October 6th, the British flew Bohr alone from Sweden to Scotland in an unpressurized aircraft. Bohr fainted during the trip. He spent the night at the home of the airport commander and the next day was flown to London. That evening he met with Sir John Anderson, the physical chemist in charge of the nascent British atomic bomb project. Anderson gave him a briefing on the Anglo-American program. According to Bohr's son Aage, who followed his father to England a week later and was his assistant throughout the war, Bohr was shocked by how far the Anglo-American program had progressed.

Bohr's alarm very likely had two sources. First, during the 1930s, when nuclear physics was developing, Bohr said on several occasions that he thought any practical use of nuclear energy was all but impossible. That view was reinforced in the spring of 1939, when he realized an important detail concerning the fission of uranium. In December 1938, the German physical chemists Otto Hahn and Fritz Strassmann had discovered that uranium could be fissioned if it was bombarded with neutrons. (Hahn's former colleague Lise Meitner and her nephew Otto Frisch, both of whom had escaped from Germany, were the first to point out that the uranium nucleus had actually split. Hahn and Strassmann were quite unclear about what they had observed. Frisch coined the term "fission" for the process.) The experiments used natural uranium, 99 percent of which is in the isotope uranium 238. About seven-tenths of a percent is in the isotope uranium 235, whose nucleus contains three fewer neutrons.

Chemically, the isotopes are indistinguishable. Bohr realized that because of their structural differences, only the very rare isotope uranium 235 had fissioned in the Hahn–Strassmann experiments. He concluded, then, that making a nuclear weapon would require the resources of an entire nation—it did, several—because it would require separating these isotopes. In December 1939, he said in a lecture "With present technical means it is, however, impossible to purify the rare isotope in sufficient quantity to realize a chain reaction." One can therefore well understand why Bohr was surprised to learn, four years later, that the Allies were doing just that.

The second reason for Bohr's alarm can be traced to his meeting with the German physicist Werner Heisenberg in mid-September 1941, almost two years before his escape to Britain. By 1941 the Germans had occupied Denmark for more than a year. During that time, they had established a so-called German Cultural Institute in Copenhagen to generate German cultural propaganda. Among other activities, the institute organized scientific meetings. Heisenberg was one of several German scientists who came to Copenhagen at its behest, in this case to a meeting of astronomers that was boycotted by the physicists from Bohr's institute. Heisenberg had known Bohr since 1922 and had spent a good deal of time at Bohr's institute, where Bohr had acted as a kind of muse for the creation of the quantum theory. Now Heisenberg had returned as a representative of a despised occupying power, touting the certainty of its victory in the war, according to some accounts.

Heisenberg spent a week in Copenhagen, visiting Bohr's institute several times. During one of these visits, he and Bohr talked privately. Neither man seems to have made any notes, so one cannot be entirely sure what was said. Also, Bohr had a reputation as a poor listener, so the two may well have talked past each other. Nevertheless, Bohr came away with the distinct impression that Heisenberg was working on nuclear weapons. As Aage Bohr later recalled "Heisenberg brought up the question of the military applications of atomic energy. My father was very reticent and expressed his skepticism because of the great technical difficulties that had to be overcome, but he had the impression that Heisenberg thought that the new possibilities could decide the outcome of the war if the war dragged on."

Now, two years later, Bohr was first learning of the Allied nuclear weapons program. What had the Germans done during those years? No wonder Bohr was alarmed.

It would be fascinating to know in detail what "new possibilities" meant, but one can make an educated guess. In 1940, a year before this visit, C.F. von Weizsäcker (who, incidentally, accompanied Heisenberg to Copenhagen and whose father was the number-two man in Hitler's Foreign Ministry) had suggested to German Army Ordnance that transuranic elements (ultimately plutonium) be used to make nuclear explosives. Plutonium is slightly heavier than uranium, has a different chemistry and is even more fissionable. Unlike uranium, though, plutonium does not exist naturally and must be manufactured in a nuclear reactor by bombarding the reactor's uranium fuel rods with neutrons. Once made, the plutonium can be chemically separated from its uranium matrix.

From the moment this process was understood, any reactor became, in a sense, a component of a nuclear weapon. There is no doubt that Heisenberg knew this well when he visited Bohr. He even gave lectures, whose texts have been preserved, describing such a possibility to highly placed Nazi officials. Is this what Heisenberg meant? He did not, it appears, discuss technical details with Bohr. As far as I can learn, Bohr did not understand the plutonium alternative until he came to Los Alamos, which means that Heisenberg did not tell Bohr about it. But what was he trying to tell Bohr? There was such a lack of agreement between what the two men said and why, we will probably never know for sure.

As a corollary to this larger puzzle there is a smaller one. Let me describe how I learned of it. Beginning in November 1977, I conducted a series of tape-recorded interviews with the physicist Hans Bethe. These sessions spanned two years and resulted in a three-part profile for *The New Yorker* magazine and a subsequent book. The interviews followed the chronology of Bethe's life. Bethe, who was born in Strasbourg in 1906 and is part Jewish, was forced to emigrate from

Germany. Bethe came to the United States in 1935 and has been at Cornell University ever since. Bethe became an American citizen in 1941, by which time, as he recalled, he was "desperate to make some contribution to the war effort." Initially, like Bohr, he was certain that nuclear weapons were entirely impractical and went to work on the development of radar at the Massachusetts Institute of Technology.

In the summer of 1942, J. Robert Oppenheimer convened a study group at the University of California at Berkeley to investigate nuclear weapons. By this time, Bethe was acknowledged as one of the leading nuclear theorists in the world, so Oppenheimer naturally asked him to participate. On the way to California by train, Bethe stopped in Chicago to pick up Edward Teller. There Bethe got a chance to see Enrico Fermi's developing nuclear reactor and, in his words, "became convinced that the atomic bomb project was real, and that it would probably work." He spent that summer working on the theory of nuclear weapons and in April 1943 went to Los Alamos, which had just opened as a laboratory. Eventually he became head of its theory division.

Now back to Bohr. On November 29, 1943, Bohr and his son Aage sailed from Glasgow on the *Aquitania* for New York City. They arrived on December 6th. Bohr was assigned the code name Nicholas Baker and Aage became James Baker; they were also given bodyguards. On December 28th, after high-level meetings in Washington, D.C., with many officials—including Major General Leslie R. Groves, the commanding officer in charge of the Manhattan Project—Bohr departed for Los Alamos. On the 31st, presumably just after arriving at the laboratory, he met with a select group of physicists. They included, in addition to Bethe, Oppenheimer and Aage Bohr, Victor Weisskopf, Edward Teller and Robert Serber. Robert Bacher, Oppenheimer's deputy, attended the latter part of the meeting when Oppenheimer had to be absent. The purpose of the meeting, which seems to have been inspired by Bohr's earlier conversations with Groves, was to discuss any new information that Bohr might have about the German nuclear program. It does not seem that any notes were taken of this meeting either—at least none have come to my attention—so one has to rely on the memories of the people who attended it.

The first time I heard of this meeting was in one of my interviews with Bethe. It came up almost casually when I asked him what the scientists at Los Alamos knew about the German nuclear program. In the course of his answer he told me about the meeting and a drawing. This is what he said (I have it on my tapes): "Heisenberg gave Bohr a drawing. This drawing was transmitted by Bohr later on to us at Los Alamos. This drawing was clearly the drawing of a reactor. But our conclusion was, when seeing it, these Germans are totally crazy. Do they want to throw a reactor down on London?" Only after the war did the Los Alamos scientists learn that the Germans knew perfectly well, at least in principle, what to do with a reactor—use it to make plutonium. But Bohr was concerned that this kind of reactor could actually be used as some sort of weapon. Furthermore, taking Bethe's remarks on their face, it would appear that Heisenberg actually gave Bohr a drawing of this reactor at their meeting in Copenhagen.

As far as I know, until I described this matter in *The New Yorker*, no one had ever mentioned such a drawing in print. In fact, my article on Bethe was often cited as the source of this odd sidelight to the Bohr–Heisenberg relationship. I found myself a kind of footnote to a footnote to history. This is where things stood until early 1994, when my authority was shaken. It happened during one of my periodic visits to the Rockefeller University in New York City, where I am an adjunct professor. Abraham Pais, a biographer of both Einstein and Bohr and a professor of physics emeritus at the university, called me into his office. I have known Pais for 40 years but had not seen him for awhile.

This was his first opportunity to tell me about a call he had received several months earlier.

It was from Thomas Powers, who at the time was writing his book *Heisenberg's War.* Powers had learned about the drawing from my book on Bethe. He was struck by the fact that at first glance it seemed as if Heisenberg had given to Bohr, in the middle of a war, a drawing of a highly classified German military project. That was such an extraordinary thing for Heisenberg to have done, if indeed he did do it, that Powers wanted to check the matter out. (Incidentally, one thing that Heisenberg certainly did *not* transmit to Bohr was the Germans' knowledge of plutonium—something much more significant than any sketch of a reactor.) He got in touch with Aage Bohr in Copenhagen (his father had died in 1962). In a letter dated November 16, 1989, Aage Bohr wrote "Heisenberg certainly drew no sketch of a reactor during his visit in 1941. The operation of a reactor was not discussed at all."

Stunned, Powers next contacted Bethe, who repeated exactly what he had told me 10 years earlier. In a quandary, Powers had called Pais, and now Pais was asking me. But Pais had done his own investigation. He had spoken with Aage Bohr, who once again insisted there had never been such a drawing. Pais also checked the archives in Copenhagen where all Bohr's private papers and journals are stored. Nowhere, he told me, did he find any mention of it. Now it was my turn to be stunned. It is one thing to be a footnote to a footnote to history, but it is quite another to be a footnote to a footnote to incorrect history.

I promised Pais I would look into the matter myself, although, in truth, when I left his office I had not the foggiest idea how I would go about it. Obviously, contacting Bethe again would not get me much further. Nothing could be more direct than what he had told me and repeated to Powers. I would need witnesses independent of Bethe and

Aage Bohr. That much was clear. But who? Oppenheimer was dead. Niels Bohr was dead. Who else could have seen that drawing?

I began, in fact, with less information than I have so far given the reader. All Bethe had told me was that Bohr had "transmitted" a drawing to Los Alamos. I thought that somehow the drawing was brought there by a courier of some kind. Bethe had not told me any details about the meeting, so initially I didn't know who might have been there. I did not even have the exact date. All that, I learned subsequently. But I did know physicists who were at Los Alamos at the time and who might have seen or heard about the drawing. Two came to mind. One was Victor Weisskopf, an old friend who had been close to Oppenheimer.

The other was the late Rudolf Peierls, who died in the fall of 1995. Peierls and Otto Frisch had, in March 1940, made the first correct calculation to determine in principle the amount of uranium 235—the so-called critical mass—needed to make a bomb. (That this mass turned out to be pounds rather than tons is what really prompted the Allied effort, not the letters sent to Franklin Roosevelt that Einstein signed.) Peierls, along with Frisch, was at Los Alamos as of early 1944. I had known Peierls for many years and we had frequently discussed the history of nuclear weapons. So it was quite natural for me to write him. This I did in early February 1994.

Soon after, both men answered. Peierls replied that he had never seen the "famous sketch," yet he did not think either Bethe or Aage Bohr had deliberately lied. He suggested that perhaps Niels Bohr had kept knowledge of the sensitive document from his family or that perhaps Heisenberg had only shown a sketch to Bohr, who might then have redrawn it. He proposed that I contact Bethe about this possibility. Weisskopf also wrote, proposing I contact Bethe once more because he too had never seen or heard about the drawing.

Neither letter was what I had hoped to receive. Weisskopf's was especially disturbing because, by this time, I had learned that he had been at the meeting where the drawing had allegedly been discussed. Clearly, I had to write Bethe, to tell him what I had learned and to see if he could shed any further light on the situation. But then I had an inspiration. I would call Robert Serber, an old friend and a professor of physics emeritus at Columbia University, who lives in New York City. After receiving his Ph.D. in 1934 from the University of Wisconsin, Serber had won one of five National Research Council Fellowships in physics and chose to go to work with Oppenheimer at Berkeley. During the next few years, they had become very close.

After a brief interlude at the University of Illinois from 1938 to 1942, Serber returned to Berkeley to work on the bomb with Oppenheimer. He was there in the summer of 1942 when Bethe and Teller arrived. By March 1943 he had moved, with the first batch of scientists, to Los Alamos. One of his early tasks was to give a series of introductory lectures on bomb physics to the arriving scientists. These were collected into what became *The Los Alamos Primer*, declassified in 1965 and first published in its entirety in 1992. If anyone would know about the drawing it would be Serber, because he was in constant contact with Oppenheimer throughout this period.

I called Serber and immediately knew that I had struck a gold mine. Not only did he remember the drawing vividly, but he also remembered the precise circumstances under which he had seen it. He had been summoned to Oppenheimer's office on December 31st, where the meeting was already in progress. Oppenheimer showed him a drawing with no explanation and asked him to identify it. (This was the kind of intellectual game Oppenheimer liked to play.) Serber looked at it and said it was clearly a drawing of a reactor. He also thought its design looked a little "silly." Oppenheimer, Serber remembers distinctly, said that in fact it was a drawing of Heisenberg's reactor and had been given to the assembled group by Bohr. Bohr, who was, as Serber remembered it, standing next to Oppenheimer, did not disagree. When I asked Serber if he remembered enough about the drawing to re-sketch it, he said he did not.

This is what Serber told me. But he also said he had some written material related to this meeting. A few days later, copies of two documents arrived: a letter from Oppenheimer to General Groves sent the day after the meeting, and a two-page memorandum written by Bethe and Teller on the explosive potential of the reactor. Unfortunately, although these documents were very suggestive, they did not, at least when I first read them, settle the issue completely. The Beth–Teller memorandum did hold significant clues, but I will return to them later. Oppenheimer's letter made no mention of the drawing, Heisenberg or the Germans. (This is not too surprising since, as Bethe had told me, information on the German effort was one of the most tightly held secrets at Los Alamos.) The last sentence of the letter read "The purpose of the enclosed memorandum is to give you a formal assurance, together with the reasons thereof, that the arrangement suggested to you by Baker [Bohr] would be a quite useless military weapon." This shows that Bohr had met with Groves before coming to Los Alamos and suggests that the "arrangement" was Bohr's rendering of a German design. Why else would Oppenheimer have to give Groves "formal assurance" that it would be a "quite useless military weapon"? Indeed, if it was not Bohr's rendering of a possible German weapon, why did Bethe and Teller stop everything they were doing to analyze it in an urgent manner?

Meanwhile, I had at last written again to Bethe, and on March 2nd, I received his answer. It begins: "I am quite positive there was a drawing. Niels Bohr presented it to us, and both Teller and I immediately said, 'This is a drawing of a reactor, not of a bomb'. . . Whether the drawing was actually due to Heisenberg, or was made by Bohr from memory, I cannot tell. But the meeting on 31 December was especially called to show us what Niels Bohr knew about the Germans' idea of a bomb."

Bethe offered a theory to explain the mystery: "Heisenberg thought that the main step to a bomb was to get a reactor and to make plutonium. A reactor, however, could also be used as a power source. Niels Bohr was very ignorant about the whole subject. Heisenberg probably wanted to show Bohr that the Germans were not making a bomb but merely a reactor. Bohr misunderstood completely, and only on 31 December 1943 was it finally explained to him that this was not a bomb. That drawing made a great impression on me. Again, I am surprised that Viki [Weisskopf] and Aage have forgotten about it. What does Serber say?"

I was able to write Bethe and tell him what Serber had said. I also wrote Teller to ask for his recollections of the meeting. I was not sure I would get an answer, and never have. In addition, I spoke by telephone to Professor Bacher. He recalled the meeting and its purpose. He did not recall the drawing, but noted that he had come late, after Oppenheimer had left, and that the drawing might have been discussed earlier. I also wrote again to Weisskopf, sending him copies of the memorandums from Serber. On February 23rd, I received a typically gracious Weisskopf letter, acknowledging that he had indeed seen the sketch but had later forgotten about it.

I now thought I had enough material to return to Pais. I played my Bethe tape and gave him copies of all the documents. He was about to return to Copenhagen, where he and his Danish wife spend about half the year, and promised me that he would speak to Aage Bohr at an opportune moment. That happened in late June. By the

30th, Pais had written to tell me what had happened. He and Aage Bohr had met, discussed the letters and reviewed the tapes. Still, Aage Bohr felt certain that Heisenberg never gave any such drawing to his father. So I wrote to him directly. In February 1995, Finn Aaserud, who is helping to edit Bohr's collected works and is in constant contact with Aage Bohr, wrote, "Aage Bohr maintains that it is entirely impossible that Bohr brought with him to the U.S. a drawing from the 1941 meeting with Heisenberg and indeed the discussion at Los Alamos you refer to had anything to do with the 1941 encounter at all."

Where does this leave us? I have asked myself this question many times since receiving both Pais's and Finn Aaserud's letters from Copenhagen. I was at a loss until, once again, I took a look at the memorandum that Bethe and Teller had prepared for Oppenheimer, Bohr and Groves. It suddenly struck me that in the first sentence of this report Heisenberg's imprint stands out like a sore thumb. It reads "The proposed pile [reactor] consists of uranium sheets immersed into heavy water." In other words, Bethe and Teller had not been considering any old reactor design but rather a very particular design that Bohr described to them. This design is actually the faulty reactor Heisenberg invented in late 1939 and early 1940, which he clung to until nearly the end of the war.

By late 1943, nearly everyone in the German program, with the exception of Heisenberg, had become convinced that uranium *plates* were inferior to a design using uranium distributed in rods or cubes. It is quite unbelievable to me that in the few short weeks between Bohr's learning about the Allied project and his arrival at Los Alamos he would have produced his own design, possessing the same flaws as Heisenberg's. It is overwhelmingly plausible that this design had come from someone connected with Heisenberg or his group. Where else could it have come from?

Let me explain in more detail. Any reactor requires fuel elements—uranium. It also requires what is known as a "moderator." This is a device that slows down the speed of the neutrons that will eventually fission the uranium. Neutrons traveling near the speed of sound—"slow" neutrons—are vastly more effective in causing fission than are the fast-moving neutrons released by the uranium nucleus when it fissions. So the fuel elements are embedded in the moderator. But a reactor designer must carefully choose the material the moderator is made from and also how the fuel elements should be placed in it. The latter involves both art and science.

The problem is that the uranium itself can absorb neutrons without producing fission. The absorption becomes greater as the neutrons are slowing down. If the geometry of the fuel elements is not well thought out, the uranium will absorb so many neutrons that a self-sustaining chain reaction will never take place. In fact, the most efficient design involves separated lumps of uranium embedded in a lattice within the moderator. (This is also the conclusion the Germans— except Heisenberg—had come to.) How big these lumps should be, and how they should be arranged, involves art. But the worst possible solution is placing uranium in sheets or layers.

In their memorandum, Bethe and Teller wrote "The proposed pile consists of uranium sheets." Heisenberg chose this design because its simpler geometry lent itself to more straightforward calculation than did the other schemes. This is the sort of thing a theoretical physicist—as opposed to an engineer—might do. Then there is the question of the moderator. Bethe and Teller state that the sheets were to be "immersed into heavy water." This specification, once explained, also has Heisenberg written all over it. The role of the moderator, as I have mentioned, is to slow down the fissioned neutrons. The best materials for this purpose are the lightest because a collision between a neutron and an object having a similar mass results in the greatest energy loss. If the neutron collides with a heavier object it will bounce off and change its direction but not its speed.

If mass were the only consideration, the ideal moderator would be hydrogen, whose nucleus is a single proton having a mass sensibly the same as the neutron's. But in reality, ordinary hydrogen fails as a moderator because it absorbs neutrons. In contrast, "heavy hydrogen," which has an extra neutron in its nucleus, does not absorb neutrons. Heavy hydrogen is found in "heavy water." But in seawater, say, this heavy water is only about one part in 5,000. So to be usable as a moderator, it must be separated from ordinary water, an expensive and difficult process.

Carbon, on the other hand, is abundant and cheap, although somewhat less effective as a moderator. By late 1940, Heisenberg had concluded that only carbon and heavy hydrogen should be used as moderators. But in January 1941, Walther Bothe, the leading experimental physicist left in Germany, began working with graphite, the form of carbon commonly used in pencils. His experiments seemed to show that graphite absorbed too many neutrons to serve as an effective moderator. What Bothe did not realize was that unless graphite is purified far beyond any ordinary industrial requirement, it will contain boron impurities. Boron soaks up neutrons like a sponge. One part boron in 500,000 of graphite can ruin the graphite as a moderator. All the same, because of Bothe's experiment, Heisenberg and the other German physicists decided that heavy water was the only practical choice for a moderator.

Needless to say, physicists who were responsible for the successful reactor program in the United States made the same kinds of calculations. Like Heisenberg, they decided that a carbon reactor would need more natural uranium than a heavy-water reactor. Fermi and his colleague Leo Szilard had also done experiments on neutron absorption in carbon. The results seemed more promising and were kept secret from the Germans. Also Szilard was a fanatic about the purity of the graphite, so their's, unlike Bothe's, worked well as a moderator. Because carbon was so much more readily available, they decided to use it. Fermi's reactor, which began operating on December 2, 1942, had a lattice of uranium lumps embedded in carbon. All the German reactors, none of which ever operated, used heavy-water moderators. Look again at the first sentence of the Bethe-Teller memorandum: "The proposed pile consists of uranium sheets immersed into heavy water." It is though someone had written "Made in Germany" on this design.

Putting everything together, it seems there is little doubt that Bohr had heard enough about the German project to realize that at its core was a nuclear device that involved immersing uranium plates in heavy water. I am prepared to believe that Heisenberg did not discuss this matter in his meeting with Bohr in 1941 as Aage Bohr insists. I am also prepared to believe that the drawing the people in Los Alamos saw on the last day of 1943 was not in Heisenberg's hand. However, I am persuaded that it was a drawing of Heisenberg's reactor and that Bohr thought it was a potential weapon. It was only after the meeting in Oppenheimer's office that Bohr was persuaded that it was not.

———

This article appeared in a slightly different form in the May 1995 issue of Scientific American. *After its publication, I continued a correspondence with Finn Aaserud, and indirectly with Aage Bohr, that has shed some additional light on this matter. There are still some disagreements between us, but they have steadily narrowed. While Aage Bohr does not remember the drawing, he has no particular reason to doubt the recollections of the others. Aage Bohr thinks his father may* well have made the same sketch for General Groves. The question is, what did the sketch depict? It is Aage Bohr's view that it was a drawing of a potential explosive device of his father's design. It is my view that it was a device inspired by the German program.

Apart from the fact that the meeting at Los Alamos was called—as all the participants, including Aage Bohr, agree—to discuss what Bohr knew about the German program, there is another piece of important evidence that I did not cite in my article. During 1943, while Bohr was still in Copenhagen, he managed, by underground couriers, to have a brief correspondence with the British nuclear physicist James Chadwick. It was Chadwick who had discovered the neutron in 1932 and was now engaged in the British nuclear weapons program. Chadwick's opening letter to Bohr is dated January 25, 1943. It extends an invitation to come to England, an offer Bohr accepted the following September.

In Bohr's February response, he reiterated his view that "in spite of all future prospects any immediate use of the latest marvelous discoveries of atomic physics is impracticable. . . ." For this reason, Bohr tells Chadwick, he has decided to remain in Denmark "to help to resist the threat against the freedom of our institutions. . . ." But by the spring (the version of the second letter I have is not dated) Bohr changed his mind. He then wrote again to Chadwick. It is this letter that I will now quote in full and then analyze. It says:

> In view of the rumours going round the world, that large scale preparations are being made for the production of metallic Uranium and heavy water to be used in atom bombs, I wish to modify my statement as regards the impracticability of an immediate use of the discoveries in nuclear physics. Taking for granted that it is impossible to separate U-isotopes in sufficient amount, any use of the natural isotopic mixture would as well-known depend on the possibility to retard the fission neutrons to such a

degree that their effect on the rare U,235 isotope exceeds neutron capture in the isotope U,238 [In short, to exploit natural uranium, a moderator is needed.] Although it might not be excluded to obtain this result in a mixture of Uranium and Deuterium, there will surely be a limit to the degree to which atomic energy could be released in such a mixture, due to the decrease of the retarding effect with rising temperatures. [Presumably what Bohr had in mind here was that when the fuel heats up, the moderator becomes less effective. Heisenberg actually claimed that this would stabilize the reactor so one would not have to worry about an explosion caused by a runaway. Since he was never able to build a reactor, he did not have to deal with the consequences of this fallacious idea.] This limit would seem to be set by a temperature of the D corresponding to about 1 Volt pro [sic] atom, and this would, therefore, be the limit of the explosion power of a thorough mixture. If, however, as often suggested [By whom? This, as I will discuss below, is the crucial question.], solid pieces of U are placed in a large tank of heavy water, it might be possible to obtain a far higher temperature within the U before the critical D-temperature is reached. Since, however, at least 1 pct., of the average energy set free will independently of the amount of U used never obviously be greater than about 100 times that required to heat the water to the critical temperature. Even if this is very great compared to that obtainable with ordinary chemical explosives, it would in view of the large scale bombing already achieved hardly be responsible to rely on the effect of a single bomb of this type procurable only with an enormous effort. The situation, however, is of course quite different, if it is true that enough heavy water can be made to manufacture a large number of eventual atom-bombs, and although I am convinced that the arguments here outlined are familiar to experts, I hasten therefore to modify my statement.

One can only wonder what Chadwick made of this communication. I can only say what I would have made of it, especially informed in hindsight of what the Germans were actually doing. The first thing that would have struck me was the phrase "In view of the rumours going round the world. . . ." I would have asked, What rumors? Whose rumors? How has Bohr been getting his information? We do know that in March 1943 Swedish radio broadcast news of the sabotage of the heavy water plant in Norway that had been supplying the Germans. We also know that the German press had been describing rumors of a "new bomb," which used the principle of "atomic destruction." Are these the rumors to which Bohr was referring?

The second thing that would have struck me is that the substance of the rumors cited by Bohr—"that large scale preparations are being made for the production of metallic Uranium and heavy water. . ."—had nothing whatsoever to do with the Allied program. Heavy water is irrelevant to building an atomic bomb except as it might be used in a reactor to manufacture plutonium. (The penultimate sentence in Bohr's letter, where he speculates about purifying "enough heavy water. . . to manufacture a large number of eventual atom-bombs" shows clearly that Bohr did not at that time have a clue about what the Allies were actually doing.) In fact, apart from the manufacture of plutonium, about which Bohr was totally ignorant, the Allied program focused precisely on the separation of uranium isotopes which, in the letter, Bohr says is impossible on a large enough scale to be used in a nuclear explosive.

The "rumours" Bohr cites are in fact an excellent description of Heisenberg's program to build a reactor. At this very time, Heisenberg was directing a project, as we have seen, to immerse metallic uranium in heavy water. To me, it is extremely plausible that someone told Bohr about this between his first and second letter to Chadwick. No other explanation seems reasonable. It is Aage Bohr's contention that if this were true, then

Bohr would have said so in his letter to Chadwick. I disagree. These letters were being exchanged by underground couriers—one of whom, incidentally, inserted the microfilmed note he was carrying in the hollow a dentist made in his tooth! In view of this, is it reasonable to imagine that Bohr would have written Chadwick something that, if intercepted, would have had disastrous consequences for whoever was supplying this information (the source of the "rumours going round the world"?) The people at the meeting at Los Alamos on December 31, 1943, believe that they had been called in to analyze Heisenberg's reactor. I believe this, indeed, is what they were doing.

A Brief History of Black Holes: Einstein to Oppenheimer

One of the things that great art and great science have in common is that the legacy of the practitioners sometimes outstrips not only their imaginations, but also their intentions. A case in point is provided by the history of black holes and, above all, by the role of Albert Einstein. In 1939, Einstein published a paper in the journal *Annals of Mathematics* with the daunting title "On a Stationary System with Spherical Symmetry Consisting of Many Gravitating Masses"—a paper which, incidentally, like all of Einstein's papers after he emigrated here in 1933, was in English. In this paper, Einstein sought to prove that "black holes," as they came to be known three decades later, were impossible. The irony is that his argument made use of his own general theory of relativity and gravitation published in 1916, the very theory that is now used to argue that, not only are black holes possible, they are for many astronomical objects, inevitable. One of the purposes of this essay is to describe Einstein's arguments and why they are irrelevant. Indeed, a few months after Einstein's paper appeared—and with no reference to it—J. Robert Oppenheimer and his student Hartland Snyder published a paper, one of the classics of modern astrophysics, entitled "On Continued Gravitational Contraction." They used Einstein's general theory of relativity to show, for the first time in the context of modern physics, how black holes could form and why they are black.

Any reader who has been exposed to the popular science of black holes is probably aware that—willy-nilly—this is part of the scientific legacy of Einstein's great work on the theory of gravitation. The attraction of gravity is, after all, what forms a black hole. But, it is much less likely that such a reader is aware that this entire subject is also built on a completely different aspect of Einstein's legacy—his development of quantum-statistical mechanics. Without the effects of quantum statistics, every astronomical object would eventually collapse into a black hole. Such a universe would bear no resemblance to the one we actually live in.

Einstein was inspired by a letter he received in June 1924 from a then totally unknown young Bengali physicist named Satyendra Nath Bose. Along with Bose's letter came a manuscript which had already been rejected by one British scientific publication. After reading the manuscript, Einstein translated it into German and arranged to have it published in the prestigious journal *Zeitschrift für Physik*.

Why did Einstein think this manuscript was so important and where did it lead? For two decades Einstein had been struggling with the nature of electromagnetic radiation, especially the radiation that is trapped inside a heated container and comes into equilibrium with—that is, acquires the same temperature as—its walls. At the turn of the century, the German physicist Max Planck had discovered the mathematical function that describes how the various colors—wavelengths—of this so-called "black body" radiation are distributed in intensity. It turns out that the form of this distribution does not depend on the material of the container walls. What matters is the temperature of the radiation. One of the most striking examples of black body radiation is the radiation—photons—left over from the Big Bang. These form a nearly perfect black body spectrum characterized by a temperature that, as we mentioned, was recently measured to be 2.726 ± 0.002 degrees above absolute zero. In this case the entire universe is the "container."

In 1905, the same year he formulated the theory of relativity, Einstein wrote a paper on black body radiation. It was for this work that he won the 1921 Nobel Prize in physics. Einstein did not attempt to derive Planck's formula—that required quantum statistics—but he showed that if one assumed the Planck law, then, under certain circumstances, light resembles a collection of energetic massless particles. What had caught Einstein's attention in Bose's manuscript was that Bose had somewhat serendipitously shown how to do the statisical mechanics of such an ensemble, leading to his derivation of the Planck law. But Einstein took the matter a step further. He used the same methods to examine the statistical mechanics of massive particles of a gas obeying the same kind of counting rules that Bose had used for the photons. He derived the analog of the Planck law for this case and noticed something remarkable. If one lowers the temperature of a gas, then at a certain critical temperature, which depends on the nature of the gas, all the molecules suddenly collect into a single degenerate state—something that is now known as "Bose–Einstein condensation" (although Bose had nothing to do with it). A case in point is a gas made up of the common helium isotope, whose nuclei consist of two protons and two neutrons. At a temperature of 2.18 degrees above absolute zero this gas turns into a liquid, which has the most uncanny properties imaginable, including a frictionless flow called superfluidity.

However, not all the particles in nature show this condensation. It is the quantum statistics of these other particles that brings us to our subject. In 1925, just after Einstein published his papers on the condensation, the Austrian-born physicist Wolfgang Pauli published what has become known as the "Pauli exclusion principle, which he had discovered a year earlier." He identified a second class of particles which includes the electron, proton and neutron and had the following property: No two such identical particles—two electrons, for example— can ever be in exactly the same quantum-mechanical state. That is the "exclusion." In 1926, Enrico Fermi and Paul A.M. Dirac developed the quantum statistics of these particles—the analogue of the Bose–Einstein distribution.

The Pauli principle holds that the last thing these particles do at low temperatures is to condense. In fact, they exhibit just the opposite tendency. If you compress, say, a gas of electrons, cooling it to very low temperatures and shrinking its volume, then the electrons are forced to invade one another's space. However, the Pauli principle forbids this, so the electrons dart away from one another at speeds approaching that of light. But a gas that consists of an ensemble of rapidly moving particles will exert a pressure. For electrons and the other Pauli particles, this pressure—the so-called "degeneracy pressure"— persists even if the gas is cooled to absolute zero. It has nothing to do with the fact that the electrons repel one another electrically. Neutrons, which have no charge, do the same thing. It is pure quantum physics. But what has it got to do with the stars?

Prior to the turn of the 20th century, astronomers had begun to identify a class of very peculiar stars. A typical example is a star that circulates around Sirius—the brightest star in the heavens. This companion star has a mass about that of the Sun, but it radiates only about 1/360th of the Sun's light. To add to the intrigue, the astronomer W. S. Adams showed in 1914 that the spectrum of these stars was very similar to any normal star. The only way to reconcile this with their dimness was to assume that the dimensions of this star were tiny compared with a normal star. In fact, the companion of Sirius has a radius that lies between that of the Earth and the planet Uranus. This implies that it has a humongous density—some 61,000 times that of water! These bizarre objects became known as "white dwarfs." What were they? Enter Sir Arthur Eddington.

When I was beginning to study physics in the late 1940s, Eddington was a hero of mine, but for the wrong reasons. I knew nothing about his great work in astronomy and astrophysics. What I admired were his popular books, which, after I knew more about physics, seemed rather silly to me. Eddington, who died in 1944, was a neo-Kantian who believed that everything of significance in the universe could be learned by examining what went on in one's head. But starting in the late teens of this century—when Eddington led one of the two expeditions that confirmed Einstein's prediction that the Sun bends starlight—until the late 1930s—when Eddington really started going off the deep end—he was truly one of the giants in 20th-century science. He practically created the discipline that led to our first understanding of *The Internal Constitution of Stars*—the title of his classic book, published in 1926. To Eddington, white dwarfs were an affront. But he did have a liberating idea.

If white dwarfs had been even hotter, some of the electrons attached to the protons in their interior would have been stripped off by collisions—the process of ionization. In that case, the atoms would lose their "boundaries" and might be squeezed together into a small dense package. What Eddington proposed in 1924 was that the gravitational pressure squeezing the white dwarf might produce the same ionization, as if the star's temperature had been increased. This was the liberating idea. But what kept the white dwarf from collapsing altogether? Here comes Pauli to the rescue. As a few of Eddington's contemporaries noted, it must be the Fermi–Dirac degeneracy pressure!

This is where things stood until the summer of 1930 when the next hero of our story, Subrahmanyan Chandrasekhar entered the scene. It was July, and Chandrasekhar, who was 19, was aboard a ship sailing from Madras to Southampton. He had been accepted by the British physicist R.H. Fowler to study with him at Cambridge University, where Eddington was also doing work. Chandrasekhar had read Eddington's book on the stars and Fowler's book on quantum-statistical mechanics and had become fascinated by white dwarfs. To pass the time, Chandrasekhar asked himself the following question: Is there any limit to how massive a white dwarf can be before it collapses under the force of its own gravitation? In retrospect, this seems like an obvious question—indeed I would not hesitate to give a simplified version of Chandrasekhar's argument to university undergraduates. At the time, however, Chandrasekhar's result set off a revolution.

In a white dwarf the electrons, which move at speeds comparable to that of light, provide the degeneracy pressure. But to keep the white dwarf electrically neutral, each electron has a compensating proton which is some 2,000 times more massive. It is the protons and the electrically neutral neutrons which are present in about the same number that supply the bulk of the gravitational compression. For the white dwarf not to collapse, the two pressures must balance. This, it turns out, puts a limit on the number of protons and neutrons, and, hence the mass of the white dwarf. This limit is known appropriately as the "Chandrasekhar limit" and is about 1.4 times that of the mass of the Sun. Any white dwarf more massive than this cannot be stable.

Then what? When Eddington learned of Chandrasekhar's result, this is just the question he asked and he was not pleased with the answer. Unless some mechanism could be found for limiting the mass of any star that was eventually going to compress itself into a white dwarf, or unless Chandrasekhar's result was wrong, massive stars were fated to gravitationally collapse into oblivion. Eddington found this intolerable and he proceeded to attack Chandrasekhar's use of quantum statistics both publicly and privately. Chandrasekhar was devastated but he held his ground, bolstered by people like Niels Bohr who assured him that Eddington was simply wrong and should be ignored. In the meantime, Chandrasekhar began studying the masses of known white dwarfs and found that they all satisfied his limit.

The stage is now nearly set for Einstein, Oppenheimer and Snyder. But first we must return to Einstein's theory of gravitation.

As far as I know, Einstein never spent a great deal of time looking for exact solutions to his gravitational equations. In the first place, the equations are very complicated, but more importantly, Einstein never thought they were complete. The gravitational part seemed complete, but you had to put in by hand the part that described the gravitating matter. That part had all the complications of quantum theory and the rest of it. Putting this together with gravitation is a deep unsolved problem. Hence, Einstein was satisfied with approximate solutions that described, with sufficient accuracy, things like the bending of starlight. Nonetheless, he was very impressed when, in 1916, the German astronomer Karl Schwarzschild actually came up with an exact solution for a a nontrivial situation.

We give students who are first learning about Newtonian gravitation the example of a test body—a "planet"—moving under the influence of a perfect sphere of gravitating matter—a "sun." We show the students that what the test body "sees" gravitationally is equivalent to a single gravitating point with the same mass as the sphere. Then we show them that the planetary orbits are conic sections—parabolas and the like. This was the problem Schwarzschild solved for general relativity. However, in general relativity the result appears under a very different guise. According to this theory, gravity distorts the geometry of space and time. This shows up in the formulae for the distance between two "events"—points in space at different times. This distance is characterized by what is called a "metric"—a mathematical form that determines it. In general relativity, a particle moves from point to point in space and time along a curve determined from the geometry produced by the gravity. Hence, Schwarzschild's starting point was to write down the metric for this case. However, when one looks at the metric—especially in the form that Schwarzschild actually used— there is something very disturbing about it. There is a distance from the center of the sphere at which the metric appears to go berserk. At this distance, which is now known as the "Schwarzschild radius," the coefficient of time vanishes and the coefficient of space becomes infinite. The metric becomes what mathematicians call "singular."

The Schwarzschild radius is not, in general, the same as the radius of the gravitating sphere. For the Sun, for example, it is three kilometers, whereas for a spherical mass of one gram it is 10^{-28} centimeters. For those of you who like mathematical formulae it is expressed in terms of the mass M of the sphere, the velocity of light c, and Newton's constant of gravity G by the simple expression

$$2\frac{GM}{c^2},$$

from which these two numerical results follow. Schwarzschild was, of course, aware that his formula went crazy at this radius, but he decided that it didn't matter. He constructed a simplified model of a "star" and showed that it would take an infinite pressure to compress it to his radius. Hence he argued it was of no practical interest. But this result did not satisfy everybody. It bothered Einstein because Schwarzschild's model star did not satisfy the requirements of the relativity theory on which the rest of the calculation was based. However, it was shown by various people that one could rewrite Schwarzschild's metric so that it looked nonsingular.

But was it really nonsingular? It would be incorrect to say that a debate "raged" because most physicists had rather little interest in these matters—at least until 1939. In a paper that year, Einstein credited his renewed concern about the Schwarzschild radius to discussions with the Princeton cosmologist H.P. Robertson and to Peter Bergmann, who was then his assistant and is now a professor emeritus at Syracuse University. It was certainly Einstein's intention in this paper to kill off the Schwarzschild singularity once and for all. At the end of it he writes "The essential result of this investigation is a clear understanding as to why 'Schwarzschild singularities' do not exist in physical reality." We would, as I will shortly make clear, rephrase this to say: If this result were relevant, it would give us a clear understanding of why black holes can't exist.

For his model, Einstein used a collection of small gravitating particles moving in circular orbits under the influence of one another's gravitation. The orbits were restricted this way to make the calculation possible. As he said, "This is a system resembling a spherical star cluster." He then asked whether such a configuration could collapse under its own gravity into a stable star with a radius equal to its Schwarzschild radius. He concluded that it could not, because at a somewhat larger radius, the stars in the cluster would have to move faster than light to keep the configuration stable. I was much taken by the fact that the then 60-year-old Einstein presents in this paper tables of numerical results, which he must have gotten by using a slide rule. However, the paper, like the slide rule, is now a historical curiosity.

While Einstein was doing this research, an entirely different enterprise was unfolding in California. Oppenheimer and his students were creating the modern theory of black holes. Unlike Einstein, who almost always worked alone, Oppenheimer and his students traveled in packs, migrating from Cal Tech to Berkeley with the season. Oppenheimer had a production line of extremely gifted students working on various problems. The curious thing about the black hole research is that it was inspired by an idea that turned out to be entirely wrong.

In 1932, the British experimental physicist James Chadwick discovered the neutron—the neutral component of the atomic nucleus. Very soon after, speculation began—most notably by Fritz Zwicky at Cal Tech and independently by the brilliant Soviet theoretical physicist Lev Landau—that there was a possible alternative to white dwarfs. When the gravitational pressure got large enough, an electron in a star could react with a proton to produce a neutron and an elusive neutrino, which then escaped from the star. Zwicky even conjectured that this is what would happen in supernova explosions. He was right. We now identify these "neutron stars" with pulsars. In the early 1930s, when Landau did this work, the actual mechanism for generating the energy in ordinary stars was unknown. This was discovered in 1938 by Hans Bethe and C.F. von Weizsäcker. Until their work was accepted, one idea for generating this energy was to place a neutron star nucleus at the center of ordinary stars, in somewhat the same spirit that many astrophysicists now conjecture that there is a black hole at the center of quasars, providing them with their extraordinary energy. The question then arose: What was the equivalent of the Chandrasekhar mass limit for these stars? This is much harder to determine than the limit for white dwarfs. The reason is that the neutrons interact with one another with a strong force whose details we still do not fully understand. Gravity will eventually overcome this force—whatever it is—but the precise limiting mass is sensitive to the details. Oppenheimer published two papers on this matter with his students Robert Serber and George Volkoff and concluded that the mass limit here is comparable to the Chandrasekhar limit for white dwarfs. (Actually the original calculation produced a result of about half the Chandrasekhar limit, but recent calculations push the limit nearly up to that of white dwarfs.) The first of these papers was published in 1938 and the second in 1939, by which time it was becoming clear that neutron star cores had nothing to do with stellar energy generation.

Oppenheimer then went on to ask exactly the same question that Eddington had asked about white dwarfs: What would happen if one had a collapsing star whose mass exceeded any of the limits? Einstein's concern, which Oppenheimer and his students were certainly unaware of—they were doing their work at the same time, separated by 3,000 miles—was of no relevance. Oppenheimer did not want to construct a stable star with a radius equal to its Schwarzschild radius. He wanted to see what would happen if one let the star collapse through its Schwarzschild radius. This is the problem he suggested that Snyder work out in detail. To simplify life, Oppenheimer told Snyder to neglect the effects of things like the degeneracy pressure or the possible rotation of the star. These might slow the collapse, but Oppenheimer's intuition told him that they would not change anything essential. He also told Snyder to assume that perfect spherical symmetry was maintained during the collapse. These assumptions were challenged many years later by a new generation of researchers using sophisticated high-speed computers—poor Snyder had an old-fashioned mechanical desk calculator. But Oppenheimer was right: nothing essential is changed. With the simplified assumptions, Snyder could follow the collapse of the star.

It turns out, however, that what is predicted to be observed depends dramatically on the vantage point of the observer. Let us start with an observer at rest a safe distance from the star. Let us also suppose that there is another observer attached to the star—"co-moving" with it—who can send light signals back to his stationary colleague. The stationary observer will see the signals from his moving counterpart gradually red-shifting. If the frequency of the signals is thought of as a clock, then the stationary observer will say that the moving observer's clock is gradually slowing down. Indeed, at the Schwarzschild radius, the clock will slow down to zero. The stationary observer will argue that it took an infinite amount of time for the star to collapse to its Schwarzschild radius. What happens after that we can't say since, according to the stationary observer, there is no "after." As far as this observer is concerned, the star is frozen at its Schwarzschild radius. Indeed until 1967, when the physicist John Wheeler coined the term "black hole," these objects were often referred to in the literature as "frozen stars." This is the real significance of the singularity in the Schwarzschild geometry. As Oppenheimer and Snyder observed in their paper, the collapsing star "tends to close itself off from any communication; only its gravitational field persists"— that is, a black hole has been formed.

But what about the observer riding with the star? This person, Oppenheimer and Snyder pointed out, has a completely different sense of things. To him, the Schwarzschild radius has no special significance. He passes right through it and on to the center in a matter of hours (for stars of typical mass), as measured by clocks attached to these observers. They would, however, be subject to monstrous tidal gravitational forces that would tear them to pieces.

The year this work was published was 1939, and the world was about to be torn to pieces. Oppenheimer was soon to go off to war to build the most destructive weapon ever devised by mankind. He never worked on the subject of black holes again. As far as I know, Einstein never did either. After the war, Oppenheimer became the director of the Institute for Advanced Study where Einstein was still a professor. From time to time they talked. There is no record of their ever having discussed black holes.

———

In 1967, Robert Oppenheimer died at the age of sixty-three, after a dreadful and courageous struggle against throat cancer. A year later his former student, and very close friend, Robert Serber decided to organize a conference at the Institute for Advanced Study in Princeton, of which Oppenheimer had been the director. Serber wanted to model the conference after the so-called "Shelter Island" conference of 1948. Shelter Island is a small island off Long Island in New York State. On that occasion, the few participants, representing the elite in physics, took over the Rams Head Inn. The war had been over for only a short time, but already there were monumental developments to be discussed in experimental and theoretical physics. In particular, using techniques created for radar, some experiments had revealed tiny electromagnetic effects which required the creation of the modern discipline of "quantum electrodynamics"—the most quantitatively accurate branch of science developed so far. These experiments were discussed at the conference and inspired new and important work.

One thing that was not discussed at this conference was the topic of black holes even though Oppenheimer, who was a dominant figure at the conference had, along with his student Hartland Snyder, in 1939, written the paper that ultimately led to all the subsequent developments in the subject. In 1948, general relativity was very much out of fashion and Oppenheimer was, if nothing else, a man of fashion. Indeed, the subject of black holes was still very low on the agenda at the first Oppenheimer memorial conference in Princeton twenty years later. As far as I recall—I was in attendance—there was one lecture on the subject. It was given by the Princeton physicist John Wheeler. I have a vivid recollection of the atmosphere. The lecture was set for late in the afternoon of the final day—not exactly prime time. In the audience of about thirty people were the likes of Richard Feynman, who had been Wheeler's student; Julian Schwinger, who had been my teacher; Murray Gell-Mann, who had discovered quarks; and Freeman Dyson. No one was much looking forward to this lecture. Although everyone was fond of Wheeler, the general feeling was that he had to some extent gone off into outer space. He had recently built a much publicized air-raid shelter in his backyard—large enough, he once told me, to house his family and several neighbors. The shelter was erected as protection against an attack by a hydrogen bomb, a weapon that he had played a significant role in developing. There was also some resentment on the part of his colleagues at his having siphoned off some of the most brilliant of the Princeton students to work on what seemed to many to be a backwater of problems involving general relativity and gravitation.

Wheeler began his lecture by announcing a theorem, partly proven and partly conjectured, which he called the "No hair theorem." As I remember it, the first line on whatever communication device he was using read "Black holes have no hair." This seemed to lack gravitas. Indeed, no one with the notable exception of Freeman Dyson was paying the slightest attention. Afterward Dyson remarked to me that listening to Wheeler was to him like reading Beowulf. This Delphic remark so impressed me that I proceeded to get myself a copy of Beowulf to see if it would help me to understand what Wheeler was talking about. It didn't. It was several years before I realized that what Wheeler was talking about was soon to become one of the most important developments in postwar research on black holes. Indeed, whole conferences have since been devoted to it; namely, what characterizes a black hole? What is the equivalent of the vanishing Cheshire cat's grin once the star has collapsed? The answer is next to nothing. More of the information is lost when the hole is formed. In short, black holes have no hair.

Like many an evangelist, Wheeler was at the time of this meeting a fairly recent convert. He had coined the term "black hole" in 1967 and had only become a believer a few years earlier. In fact, as late as 1958, on the occasion of one of the so-called Solvay conferences in Brussels, Belgium, Wheeler had given an address on the fate of massive stars which could have come straight from the text of Sir Arthur Eddington twenty years earlier. Eddington had decided on a priori grounds that it was absurd for a massive star to collapse into a gravitational void. He was sure that some mechanism would intervene before that would happen. Now, two decades later, here was Wheeler saying to the delegates in Brussels "Perhaps there is no final equilibrium state: this is the proposal of Oppenheimer and Snyder. . . [that is, perhaps these massive stars do collapse into black holes.]" But, he went on, "A new look at this proposal today suggests that it does not give an acceptable answer to the fate of a system of A-nucleons under gravitational forces. . . ."

In other words, Wheeler was claiming that this "new look," whatever it was, had ruled out the formation of black holes. Instead, Wheeler suggested that because of some unspecified mechanism, the "nucleons [neutrons and protons] at the centre of a highly compressed mass must necessarily dissolve away into radiation." This too sounded rather like Eddington. Oppenheimer also was at this meeting, and in the only post-

war comment I have found that he made on the subject, he asked, "Would not the simplest assumption about the fate of a star more than the critical mass be this, that it undergoes continued gravitational contraction and cuts itself off from the rest of the universe?" During the years I spent at the Institute for Advanced Study from 1957 to 1959, I never heard Oppenheimer say a single word on black holes. He was like a father who had disowned his children. He did not live long enough to see his child grow up to be a giant.

Madame Curie

Regular *contributors to the* New York Review of Books *become accustomed to the fact that its editor, Robert Silvers, often swoops down on them without warning to request that they review one or more books. This happened to me once again early in 1995, the book in question being* Marie Curie: A Life, *by Susan Quinn. Along with the book, there came a letter outlining the proposed word length. I promptly lost it—a Freudian slip? So, paying no attention to the number of words, I simply wrote my essay. Since I use a word processor, when I finished writing I asked it to count the words. There were about 11,000! Hoping for the best, I sent my review to the* Review. *After a lengthy silence, Robert Silvers called. He asked, in a tone that betrayed some concern, if I had not read his letter. I had to admit that I had lost it. He said that in it he had asked for a review of between 3,500 and 4,000 words and, instead, I had created this monster. He agreed that the subject was fascinating but insisted that major surgery was required. Between us, we managed to cut some 6,000 words from the original manuscript and, it was finally published. Now, having the luxury of space, I have restored much of the original text. I have also tried to make use of the Review's excellent editorial suggestions.*

———

Here are three quotations whose subject is Madame Curie. They are in chronological order, although I am not sure of the precise date of the middle one. The same individual is responsible for all three, but I will not reveal his name until after I have presented them. The name, I can assure you, will be a familiar one. I will give the dates and circumstances of the quotations, as far as I can determine them, when I reveal their author.

> Madame Curie is very intelligent but has the soul of a herring [*Haringseele* in the German original], which means that she is lacking in all feelings of joy and sorrow. Almost the only way in which she expresses her feeling is to rail at things she doesn't like. And she has a daughter [Irène] who is even worse— like a Grenadier [an infantryman]. The daughter is also very gifted. . . .

And:

> In Mme. Curie I can see no more than a brilliant exception. Even if there were more women scientists of like caliber they would serve as no argument against the fundamental weakness of the feminine organization.

Finally:

It was my good fortune to be linked with Mme. Curie through twenty years of sublime and unclouded friendship. I came to admire her human grandeur to an ever-growing degree. Her strength, her purity of will, her austerity toward herself, her incorruptible judgement—all these were of a kind seldom found joined in a single individual. She felt herself at every moment to be a servant of society, and her professional modesty never left any room for complacency. She was oppressed by an abiding sense for the asperities and inequities of society. This is what gave her that severe outward aspect, so easily misinterpreted by those who were not close to her—a curious severity unrelieved by an artistic strain. Once she had recognized a certain way as the right one, she pursued it without compromise and with extreme tenacity.

The greatest scientific deed of her life—proving the existence of radioactive [Madame Curie's word] elements and isolating them—owes its accomplishment not merely to bold intuition but to a devotion and tenacity in execution under the most extreme hardships imaginable, such as the history of experimental science has not often witnessed.

These quotes are taken from Albert Einstein, and indeed they span the 20 years he knew Marie Curie. They all contain a grain of truth and they all betray a lack of understanding. Let us take them one at a time. I found the first quotation, which is from Einstein's collected letters, partly quoted in Susan Quinn's biography.* The quotation dates from the summer of 1913 and is from a letter Einstein wrote to his cousin Elsa Löwenthal, whom he married six years later. At the time Einstein wrote it, he was still married to his first wife, Mileva,

*Quinn, *Marie Curie: A Life* (New York: Simon & Schuster, 1995), 351.

and was teaching at the Federal Institute of Technology in Zurich. He had met Madame Curie at the first of what became known as the Solvay Conferences—a series of meetings organized and paid for by a wealthy Belgian industrial chemist, Ernest Solvay. The first was held in Brussels in 1911, and both Einstein and Marie Curie were invited. That was the year Madame Curie won the second of her two Nobel Prizes. This one was in chemistry for her discovery of the elements she named radium and polonium. The first was in physics and had been awarded in 1903 jointly to Madame Curie, her husband, Pierre Curie, and the French physicist Henri Becquerel for their discovery of radioactivity.

In the spring of 1913, two years after this meeting, Einstein came to Paris, accompanied by his wife, to give a series of lectures. They spent an evening being entertained by Madame Curie. Apparently, plans were made for some kind of outing in the Swiss Alps— Madame Curie was a hiker and bicycle rider. It took place that summer in the Engadine, not far from Zurich by train. The group consisted of Einstein, one of his sons, Madame Curie and her two daughters. It was after this excursion that Einstein wrote to his cousin.

It is clearly not a very flattering letter. I suppose "soul of a herring" means something like "cold fish." There is no question that Madame Curie could at times appear very cold, even to her daughters. But what does it mean to be "lacking in all feelings of joy and sorrow"? Does that mean, according to Einstein, that Madame Curie felt less joy and pain than most people? If this is what he thought, he was entirely wrong. At that time, Madame Curie—and I will go into this later—was undergoing emotional upheavals that would have destroyed most people. She was a very private person and certainly would not have shared any of this with Einstein. On the other hand, I think Ms. Quinn has the purpose of Einstein's letter entirely wrong.

She writes, explaining its tone, that "Einstein was courting his cousin Elsa, who would later become his second wife, when he wrote this letter. And it seems likely that he was trying to reassure her that his hiking outing with another woman had not been any fun at all." As I mentioned before, in the summer of 1913 Einstein was living in Zurich with his wife Mileva. If anybody needed reassuring it would have been his wife Mileva and not his cousin Elsa, who was living in Berlin. Einstein moved there with Mileva the following year. They then separated and he married Elsa in 1919. I think that Einstein is simply saying in his letter that Marie Curie was no fun to hike with because she was no fun to hike with.

I came across the next quotation in a book I found in a second-hand bookstore near Harvard Square in the 1950s. The book is called *Cosmic Religion* and was published in 1931.* It is a collection of some of Einstein's aphorisms and short essays. The editor is not identified, nor are the precise dates and sources. But they clearly date from before 1931. Many of them are familiar; the one on women and Madame Curie isn't. I have never seen it quoted anywhere else. It also has a grain of truth and shows a lack of understanding. Clearly, Madame Curie was both "brilliant" and an "exception." There is brilliance and there is brilliance, but when Maria Salomea Sklodowska— her maiden name—finally got the opportunity in the 1890s to study science at the Sorbonne, she was first in her class in the physics "licence" when she took the examination in 1891 and second in her class in 1894 for the mathematics licence. This was among teachers and students of the very highest caliber. As for her being an "exception," she was one of 23 women out of over 1,800 students enrolled in the Faculté des Sciences in 1891. Finding any woman studying for

*Albert Einstein, *Cosmic Religion, with Other Opinions and Aphorisms* (New York: Covici-Friede, 1931), 105.

an advanced degree in a European university at that time was a novelty. In 1867, the University of Zurich, where Einstein's future wife Mileva matriculated in 1896, became the first European university ever to grant a Ph.D. to a woman.

But what about the second part of this quotation? First, it is certainly not surprising. Most men of Einstein's generation felt this way about women scientists and, in truth, many still do. This, of course, is another discussion, which would require another essay if not a whole book. But here at least is a suggestion as to how to begin. Imagine that Einstein had been born in Germany some 20 years earlier than his actual birth year of 1879—the time when Jews were first being allowed to study in the universities. He certainly would have been a "brilliant exception." What would the response have been if someone had written the same statement about Einstein that he wrote about Madame Curie, but substituted "Einstein" for "Curie" and "Jews" for "women"? The question almost answers itself. The issue would not have been a fundamental "weakness in the Jewish organization." It would have been a simple matter of anti-Semitism.

Nonetheless, women are still "exceptions" when it comes to science and engineering. However, we are relatively new at this. I vividly recall a faculty meeting at the engineering school where I was teaching in the late 1960s, at which the president informed the faculty that coeducation was impossible at our place because we lacked the appropriate plumbing facilities. It is amazing how rapidly an army of plumbers materialized when the same president was informed, not long afterwards, that the school would lose its federal funding if it excluded women. It has had women for some 25 years, but they still account for less than a third of the student body.

The third quotation has the feeling of an obituary, and that indeed is what it is. Marie Curie died on July 4, 1934, and this obituary was published the following year.* "Soul of a herring" has been replaced by "severe outward aspect." However, it is not clear to me whether Einstein had acquired any more understanding of Madame Curie in twenty-odd years. Obituaries like this rapidly became part of the Curie legend—replete with phrases like "profound modesty," "her purity of will," "incorruptible judgement" and the rest. This sort of thing has made every biography of Madame Curie, except Ms. Quinn's, intolerable to me. Add to this the tragic death of her husband in 1906 at age 47 and the fact that her daughter Irène Joliot-Curie, with her husband Frédéric Joliot-Curie, went on to win the 1935 Nobel Prize in chemistry, and you have the stuff of which bad movies are made. To complete the scenario, remind yourself that it was the exposure to the very radiation she had discovered that hastened Madame Curie's death. It's just too much, too awful—too good to be true. Whenever I glanced at a biography of Madame Curie, I thought of Oscar Wilde's celebrated remark that "One must have a heart of stone to read the death of Little Nell without laughing."

A few years ago, however, long before reading Ms. Quinn's book, I was jarred out of my complacency. I was giving my standard spiel, about Madame Curie being one of the most tedious people I had ever read about, to a French physicist acquaintance of mine. After listening for awhile, he interrupted to say "Your problem, my dear Bernstein, is that you do not know the slightest thing about Madame Curie." He then proceeded to give me a brief lecture, which thoroughly shook me. Until then, I did not have the remotest idea of Madame Curie's real emotional life. For example, it never would have dawned on me that, beginning in the summer of 1910 and lasting well over a year, Marie Curie had a love affair with the French physicist Paul Langevin

*One can find this fragment in A. P. French, ed., *Einstein: A Centenary Volume* (London: Heinemann, 1979), 305.

that scandalized France. Madame Curie was widowed, but Langevin was married and had four children. Intimate letters that Madame Curie had written to Langevin found their way into the tabloid press—how, I will explain later—and the scandal provoked five known duels in Paris, one of which involved Langevin himself. In fact, things got so bad that members of the Swedish Academy contacted her to try to persuade her not to accept the Nobel Prize they had just awarded her. I suspect that the ranks of Nobel Prize winners would be decimated if marital fidelity became a requirement. Einstein, one gathers, would have been one of the first to go. There is some suspicion that he had a premarital affair with his cousin Elsa while he was still married to his first wife.

I was so astonished by these revelations that I went immediately to the *Dictionary of Scientific Biography* to try to confirm them. There is a long and quite detailed article about Madame Curie. The name of Langevin does not appear. There is also a long article about Langevin. The only mention of Madame Curie is the sentence "Meanwhile he [Langevin] also taught at the École Municipale de Physique et Chimie, succeeding Pierre Curie (1904) and then at the École Nationale Supérieure des Jeunes Filles (Sèvres) replacing Marie Curie who had been widowed (1906)." The author adds "Langevin loved teaching, and he excelled at it." The myth is preserved. But let us begin at the beginning.

Maria Salomea Sklodowska was born on November 7, 1867, in Warsaw, the youngest of five children. Her mother and father were both educators. One has the impression that they supplied a very stimulating atmosphere for the children. Wladyslaw, Maria's father, while not a scientist, had a keen interest in science that he shared with his children. (There seem to be two classes of scientists: those like Marie Curie and Richard Feynman, whose interest in science was aroused by a parent—in Feynman's case, also a nonscientist father—and those like I.I. Rabi, whose interest was entirely self-generated and even went counter to the family's religious culture.) The first great tragedy of Marie Curie's life occurred when she was 10. Her mother died of tuberculosis at age 42. This was a disease, then of entirely mysterious origin, that often took many years to kill—in the case of Maria's mother, about six. Hence, Maria's early childhood, and that of the rest of the children, consisted of alternations of hope and despair as their mother tried one "cure" after another while her condition inexorably deteriorated. It is little wonder that two of Maria's siblings became doctors and that she herself devoted much of her professional life, after radioactivity had been discovered, to exploiting it for medical purposes.

While this was going on, Maria's father lost his job as the assistant director of a local school. To earn a living, the family home was turned into a small private boarding school. It was the end of a real family life for the Sklodowskas. As if this were not enough, in 1876, two years before her mother's death, Maria's oldest sister Zosia died of typhus at age 14. This was a loss the entire family felt for the rest of their lives. None of this, however, was allowed to interfere with Maria's education. It was clear to almost everybody that she was unusually gifted. She finished high school at the age of 15. Then her father, who seems to have been a man of exceptional understanding, insisted that she "drop out" for a year to live with some maternal uncles in the country. Ms. Quinn describes this time as "an unending round of all-night dances and general hilarity." It is the only year of Marie Curie's life that could remotely be characterized that way.

Upon her return to Warsaw, the family's financial realities became apparent. Her father no longer took in boarders, which meant the family's income was reduced still further. It seemed out of the question for Maria to go on to a university. For one thing, Warsaw University did not admit women, which meant going abroad. For this there was just no money. At first she did private tutoring and then took work as a governess, at age 18, with the well-off Zorawski family. But she kept studying. Ms. Quinn quotes from a letter written at the time to her cousin:

> At the moment I am reading
>
> 1. Daniel's *Physics,* of which I have finished the first volume;
> 2. Spencer's *Sociology* in French;
> 3. Paul Ber's *Lessons on Anatomy and Physiology* in Russian.
>
> I read several things at a time: the consecutive study of a single subject would wear out my poor head which is already much overworked. When I feel myself quite unable to read with profit, I work out problems of algebra or trigonometry, which allow no lapses of attention and get me back onto the right road.

While working as a governess, Maria and her older sister Bronia made a plan. Bronia would go to Paris and study medicine. After she had completed her studies she would send for Maria. Meanwhile, Maria would help support both Bronia and their father. This is what happened, but it nearly didn't. Maria and the eldest son of the Zorawskis, Kazimierz, fell in love. Indeed, they planned to get married. But when the Zorawskis heard about it, they absolutely refused to sanction the match. It must have broken Maria's heart. However, she stayed in their employ for another 15 months—partly to fulfill her financial obligations to her sister and partly, no doubt, with the hope of continuing her relationship with Kazimierz. There is an odd parallel between the outcome of this relationship and what eventually happened with Langevin. In both cases the men involved did not quite have the courage, or commitment, to make the final break with their families, and in both cases Marie Curie was left behind.

So strong were Maria's feelings about Kazimierz that in 1889, when her sister wrote that she was both finishing her studies and getting married to a fellow medical student—which meant that, after four years of waiting, Maria could finally come to Paris—Maria almost changed her mind. She was still in love with him. If she had actually married him and stayed in Poland, the history of modern physics would have been very different. As it was, she spent the next year in Warsaw looking after her father and saving money. Finally, in November 1891, at the age of 23 she left by train—fourth class—for Paris and the Sorbonne. Her sister had wanted her to move in with her and her new husband. But she found him much too gregarious, and for the rest of her student days she lived in "bachelor quarters" in various garrets.

Maria Sklodowska began signing her name "Marie" almost as soon as she arrived in Paris. One supposes that she wanted to make a clear break with the past. There was no question of what she wanted to study—science. She had the good fortune to arrive at the Sorbonne at a time when it was undergoing something of a scientific renaissance. One of her professors was Gabriel Lippmann, who won the Nobel Prize in physics in 1908. Another was Henri Poincaré, arguably the greatest mathematician of his era. These people seem to have accepted her simply as a brilliant student who happened to be a woman. After her years of self-study, the Sorbonne must have seemed like an intellectual garden salad. There were, according to her accounts, simply not enough hours in the day.

Marie's ambition upon graduation was to return to Poland and become a science teacher, presumably on the high-school level given the situation of women in the universities. But at this point—probably through professor Lippmann—she got a commission from something called the Society for the Encouragement of National Industry to study the magnetic properties of different kinds of steel. That kept her in Paris and led to her meeting Pierre Curie.

Pierre Curie was born in 1859, making him some eight years older than Marie. By the time they met in April 1894, Pierre Curie was already an established physicist with his own laboratory. He had done outstanding work on the properties of crystals and had turned his attention to the study of the magnetic properties of various substances as a function of temperature. The results of this work are still taught—with modifications—in modern physics courses, although their explanation had to await the development of the quantum theory. Had he lived longer, Pierre Curie might well have gotten a Nobel Prize for this work as well. I bring this up because in the hagiography, Pierre Curie often emerges almost as a bystander. In fact, the collaboration between Marie and Pierre Curie was a real collaboration, one in which the work would almost surely not have been done individually.

The couple was introduced by a Polish physicist named Józef Kowalski and his wife, who had met Marie earlier. They had heard about Marie's commission and the fact that she lacked adequate laboratory space to carry it out. They thought of their friend Pierre Curie who had his own laboratory and possibly extra space. Curie was himself teaching at the École Municipale de Physique et Chimie Industrielles—most definitely not one of the *Grandes Écoles*. In fact, Pierre Curie was then, and for most of his life, very definitely a nonestablishment figure. He had originally been educated at home and had never bothered to get his Ph.D. yet he had acquired a licence at the Sorbonne, after which he immediately began a program of original research with his brother. Marie was also an outsider, a foreigner in France and not quite of the right class in Poland. Marie recognized their similarities at once. She later wrote "There was between his conceptions and mine, despite the difference between our native countries, a surprising kinship, no doubt attributable to a certain likeness in the moral atmosphere in which we were both raised." Despite her sense that she might be "betraying" her country by marrying and then living outside of Poland, the attraction was too great and the couple was married on July 26, 1895, in the town hall in Sceaux. Afterwards, they went on a trip to Brittany on two new bicycles purchased for them as a wedding gift.

To understand the next step, we must review a little early history of the discovery of radioactivity. In the fall of 1895, a few months after the Curies were married, the German physicist Wilhelm Roentgen discovered what he called "X-rays." Putting Roentgen's discovery in modern—and entirely anachronistic—terms he bombarded metal plates with electrons. When these electrons collide with atoms in the metal, the atomic electrons are elevated to what much later became known as "excited states"—states of higher energy. The atomic electrons then relax back into their original "ground state" and, to conserve energy, emit the energetic electromagnetic radiation that Roentgen called X-rays. Roentgen himself had no such model in mind. Indeed the "existence" of atoms—a subject I will come back to—was still being debated. He focused on the properties of the rays, not how they were produced. The striking thing that Roentgen found was that they could readily penetrate matter. They also left a visible imprint on a photographic plate. In fact, late in December 1895 Roentgen succeeded in "photographing" the bones in his wife's left hand, thereby opening a new era in medicine. He won the first Nobel Prize in physics in 1901 for this work.

French physicist Henri Becquerel took the next step. In February 1896 he made the somewhat accidental discovery that potassium uranyl disulfate spontaneously emitted radiation. To use Madame Curie's term, it was radioactive. This was very puzzling. It appeared as if uranium was pouring out energy at a constant rate. A few years passed before it was realized that the activity fell off over time and that the rate of this falloff—which could be very slow—was characteristic of the type of atom that was decaying. Where did this energy come from? Did it fill space and somehow transform itself into the radiant energy of decay? Or was the conservation of energy itself violated? Neither Becquerel nor anyone else could say, and as a good empiricist he contented himself with the facts. Enter the Curies.

The idea of studying radioactivity seems to have been entirely Marie's. She was familiar with Becquerel's work and with the fact that it seemed to have reached a dead end. The field was open. One can imagine that it did not take much persuading to get Pierre to join the enterprise, which began on December 16, 1897, according to her lab notebooks. Pierre was a master scientific instrument builder. Their idea was to measure the degree that air became "electrified" when the radiation emitted by the decaying uranium passed through it. The term of art is "ionization." The degree of ionization would be a measure of the strength of the radioactive source. This is where Pierre's sensitive instruments for measuring electric charge came in. The first question that had occurred to Madame Curie was whether this spontaneous radioactivity was a property of uranium alone, or of other elements as well. It was here that the Curies made their first great discovery.

By the winter of 1898, Marie had tested a variety of elements for radioactivity. She found no conclusive evidence for any, apart from the uranium. This is not too surprising. She had been testing relatively light elements like gold and copper. These do have unstable types—"isotopes"—but when these elements are found in nature, the unstable isotopes have largely decayed away and one is mostly left with the stable isotope, which is not radioactive. But the heavy elements, like uranium, do not have stable isotopes. In fact they would have already decayed into lighter elements if they didn't decay so slowly. It can take over a billion years for half of any sample to decay and that is why they are still around and still decaying. This was certainly not understood when the Curies were doing this work. On the contrary, it was their discovery that started the process towards this understanding.

On February 17, 1898, Marie had what turned out be the inspired idea of testing pitchblende—a heavy black material from which uranium had first been extracted. They tested it for radioactivity and discovered, to their astonishment, that it was more active than uranium itself. What could that mean? At this point the Curies made an assumption that brought something entirely new into physics and which we have applied ever since. Since pitchblende, a very messy compound consisting of several elements, was more active than uranium, it must contain an unknown element that was emitting this radiation. From then on we have used the products of radioactivity as "fingerprints" to identify the objects that have decayed. In fact, many particles we now deal with, decay so rapidly that they can *only* be identified through their decay products. With the help of a chemist, Gustave Bémont, they attempted to isolate the unknown element. Finally, by July, after processing tons of material in vats, they were ready to announce their findings—a new element they called "polonium." As they say in their paper "If the existence of the new metal is confirmed, we propose to call it *polonium* from the name of the country of origin of one of us." By the end of the year they were able to announce the discovery of a second new radioactive element, which they named radium. They spent the next several years painstakingly measuring its properties, culminating in a 1903 paper by Pierre Curie and collaborator Albert Laborde in which they measured how much energy a gram of radium can release in an hour—enough to boil water.

To me, this was the high-water mark of their research. The torch was then passed elsewhere, especially to the great New Zealand-born experimental physicist Ernest Rutherford and his young collaborators. In preparing this essay I have re-read the early papers of the Curies and the 1902 paper of Rutherford and Frederick Soddy entitled "The Cause and Nature of Radioactivity." Reading the latter paper is like stepping into a new world—the world of modern physics. Let me state the issue. At the turn of the 20th century there was a debate as to the role of the atom. Did matter really consist, as Newton put it, of "solid, massy, impenetrable movable Particles. . .even so very hard, as never to wear out or break into pieces. . . ."* This is what one would call the *physicists'* atom. It has a mass, a shape, an electric charge and so on. Certainly, there is the implicit assumption here that if one could actually divide and re-divide matter indefinitely, one would ultimately arrive at this ineluctable component. On the other hand, there was the *chemists'* atom. This was what Einstein referred to more as a "visualizing symbol than as knowledge concerning the factual construction of matter."† In other words, one could use a model for chemical reactions that was based on the assumption that these reactions took place *as if* matter was atomic,

*This quotation is taken from Newton's *Opticks*. The relevant passage can be found in the following anthology: Henry Boorse and Lloyd Motz, eds., *The World of the Atom* (New York: Basic Books, 1966), 102. Excerpts from the papers of the Curies and Rutherford can also be found in this collection.

†See P. A. Schlipp, ed., *Albert Einstein: Philosopher-Scientist* (Evanston: The Library of Living Philosophers, 1949), 19.

without committing oneself to the "reality" of these building blocks.

No one had any difficulty with this use of the atomic hypothesis. But there was a vivid debate among scientists at this time about the "existence" of the physicists' atom. The most important skeptical voice was that of the Austrian philosopher-physicist Ernst Mach who used to ask *"Haben Sie einen gesehen?"*—"Have you seen one?" * The Curies were, as far as one can tell, firmly committed to the physicists' atom. However, they never could fully accept the idea that this atom was unstable. One can well understand their reluctance. As their own research showed, radioactivity proceeds irrespective of the state of the matter. You can heat uranium, dissolve it, or paint it blue and it will continue to decay spontaneously at the same rate. The process appears to be noncausal, indeed spontaneous. On the face of it, this was quite different from Roentgen's X-rays, which were produced by actually bombarding something. What was the mechanism that produced the spontaneous decay of atoms? That question could not even be correctly approached until the invention of quantum theory three decades later.

Then there was the question of the energy. Where did it come from? At one point Madame Curie conjectured that a radium sample might lose mass as it decays. She was right, but that is quite different from the idea that an individual atom—a building block—might lose mass in a decay, an idea the Curies had enormous difficulty accepting. But this is exactly what happens. The atomic mass-loss goes into the energy of the decay products according to Einstein's formula $E = mc^2$. However, this insight had to await Einstein's theory of relativity. Indeed in the paper where he introduced this formula in 1905 he wrote

*For a fuller discussion, especially the chapter entitled "Ernst Mach and the Quarks," see Jeremy Bernstein, *Quarks, Cranks and the Cosmos* (New York: Basic Books, 1993).

"It is not impossible that with bodies whose energy-content is variable to a high degree (e.g., with radium salts) the theory may be put to the test." Despite these problems, Rutherford had no doubt that radioactivity was an atomic phenomenon. His paper with Soddy has a concluding section that contains the sentence "Since therefore radioactivity is at once an atomic phenomenon and accompanied by chemical changes in which new types of matter are produced, these changes must be occurring within the atom, and the radioactive elements must be undergoing spontaneous transformation." He did not suggest a mechanism, nor was he overly concerned about the energy problem. His intuition told him what must be going on and he was content to wait for the theorists to catch up. This gave him an enormous advantage over the Curies. For the next few years, while Pierre was alive, they spent a lot of time trying unsuccessfully to avoid a real atomic description of radioactivity—something Madame Curie continued even after Pierre's death. Rutherford had no such burden, and he and his colleagues and students made one fundamental discovery after another.

The death of Pierre Curie on April 19, 1906, was a blow from which Marie never really recovered. Not only did their two young daughters—Irène who was born in 1897 and Eve who was born in 1904—suddenly lose their father, but Marie lost both a husband and her most intimate scientific collaborator. As I mentioned before, both Curies were outside the establishment; Marie because she was a woman and a foreigner, and Pierre, in part, by choice. In 1903, when he was proposed for the Legion of Honor he refused to allow his name to be put into nomination. The Nobel Prize was something else. When they won it in 1903—Marie being the first woman to win a Nobel and the last to win one in the sciences until 1935, when her daughter Irène shared the prize in chemistry with her husband—they didn't even attend the ceremonies. Marie was not well, but Pierre decided that he couldn't take time away from his classes! Can one imagine a present-day professor of physics writing his excuses to the Nobel Committee as Pierre did?— "We can't be gone from our classes at this time of year without incurring great difficulties in the teaching that is entrusted to us. . . ." * One's dean would come into one's office with a loaded gun in one hand, and a first-class airplane ticket to Stockholm in the other and escort one to the airport. However, by the time of Pierre's death the Curies were moving into the establishment. They had won several prizes, in addition to the Nobel, and Pierre, with great reluctance, had allowed himself to be elected to the French Academy of Sciences. Then came the disaster.

The death of Pierre Curie is such an odd amalgam of 19th- and 20th-century life that if it were presented in a novel no one would believe it. By 1906 Pierre Curie was beginning to show signs of radiation sickness. He had debilitating pains in his back and legs, and his hands were so damaged by radiation burns that he apparently had trouble dressing himself. Considering how they were working, it is astonishing to me that they weren't all dead. No present-day scientist would spend ten minutes in a laboratory as contaminated as theirs, let alone months or years at a time. No one knew then just how dangerous and insidious exposure to radioactivity was. That is the 20th-century aspect of Pierre's death. But the proximate cause of his death is right out of the 19th century. He was run over by a very heavy, 30-foot, horse-drawn wagon filled with military uniforms. His

*Quinn, Marie Curie: A Life, 192.

skull was crushed under one of the wheels, killing him instantly. He had been returning from a meeting and, it appears absent-mindedly, stepped into the street in the path of the wagon. Had he been in better health—more agile—he might have gotten out of the way.

Marie Curie was told soon afterwards. One cannot read her journal entry for that day without feeling her pain. It is almost unbearable. She writes:

> I enter the room. Someone says: "He is dead." Can one comprehend such words? Pierre is dead, he who I had seen leave looking fine this morning, he who I expected to press in my arms this evening. I will only see him dead and it's over forever. I repeat your name again and always "Pierre, Pierre, my Pierre." Alas, that doesn't make him come back, he is gone forever, leaving me nothing but desolation and despair. *

"Nothing but desolation and despair"—a weaker person would have been destroyed. But Marie Curie was one of the strongest women who ever lived. This is evident both by what she did, and how she appeared publicly. She almost never mentioned Pierre's name again. She did not want to share her grief. But privately she addressed him in her diary as if he were still alive and would read the entries. "Did you say it then? I don't remember how many times have you said to me, my Pierre: 'We really have the same way of seeing everything.'"†
In photographs from this period she looks like an eagle. It is a gaze that must have made many people uncomfortable. Einstein, who met her a few years later, was certainly one of them. It is not a gaze that would make one suspect that she is about to enter into a romantic relationship that will make her the scandal of Paris. Enter Paul Langevin.

*Quinn, *Marie Curie: A Life*, 234.
†Ibid. 237.

Langevin was born in Paris in 1872, making him five years younger than Marie Curie. Unlike Pierre Curie he went to all the right schools and did brilliantly. It seems as if he first encountered Pierre Curie when at age 17 he came to study under him at the École Municipale de Physique et Chimie. Afterwards he went to the Sorbonne and then placed first in the entrance examination for the École Normale. Those are the best academic addresses, the ones that assure a young French scientist that he or she will rise through the system like a helium balloon. By 1902 Langevin held a joint appointment at the College de France and at the École Municipale, where he replaced Pierre Curie who went to the Sorbonne in 1904. After Pierre died, Marie took over his post at the Sorbonne, and Langevin took over hers at the École Municipale Supérieure des Jeunes Filles in Sèvres. Langevin was, by all accounts, a marvelous teacher. He did not publish often, but he had the respect of physicists like Einstein who claimed that Langevin would have discovered the theory of relativity if he had not.

Before I discuss the turn that his long-standing friendship took with Madame Curie in 1910, let me make a remark about Ms. Quinn's treatment of Langevin. She leaves out half of his life. Langevin lived until 1946, long enough to see his daughter Hélène—who was born in 1909 just before his affair with Marie Curie—returned from Auschwitz. She had been in the Resistance. Yet, this is almost the last thing Ms. Quinn says about him. She writes "By 1914, according to his son André, Paul and Jeanne [his wife] were back together. Later on with his wife's acquiescence, Langevin had another mistress. But this time he chose a woman of the acceptable kind: she was an anonymous secretary." Ms. Quinn chose to read the facts of the affair almost as if Marie Curie was the victim and Langevin the aggressor. I do not see Marie Curie as anybody's victim. She knew what she wanted, and what she wanted was Langevin. The problem was that Langevin had been married since 1898 and had already fathered four children.

One can imagine what must have happened. Langevin must have been one of the few people—certainly one of the few men—with whom Marie Curie could share her real feelings about the loss of Pierre. Having been in situations like this myself, I know that it can give one a power—often a very reluctant power—over someone else that requires enormous tact and understanding to deal with. In Langevin's case, there was the fact that from its beginning his marriage had been a very stormy one. There was talk of divorce, but at the same time the couple continued having children. Between Langevin and Madame Curie there must have been a barter of condolences; his for the death of her husband and hers for what she saw, or hoped to see—the death of his marriage. In the summer of 1910 she wrote him from her vacation home—Madame Curie acquired several homes in her lifetime—by the seashore:

> My dear Paul, I spent yesterday evening and night thinking of you, of the hours that we have spent together and of which I have kept a delicious memory. I still see your good and tender eyes, your charming smile, and I think only of the moment when I will find again all the sweetness of your presence.*

That was the summer the two became lovers. In fact, they rented an apartment together near the Sorbonne where they could be alone. What Ms. Quinn does not tell us—perhaps no one knows—is whose idea this was. To put it crassly, who paid the rent? After all, Langevin was a relatively unknown academic with a wife and four children to support. Did he have the money to go around

*Quinn, Marie Curie: A Life, 265.

120

renting spare Parisian apartments? On the other hand, Madame Curie was by this time a world-renowned scientist—a Nobel Prize winner—and now the administrator of a very large laboratory. Her financial resources were becoming commensurate with her stature. Who was keeping whom? It gives one pause for thought. It certainly gave Langevin's wife pause for thought. She realized very quickly that her husband's relationship with Madame Curie had now turned into something else. She was determined to break up this happy arrangement by all means, fair or foul. Ironically, it was Marie Curie herself who supplied the necessary ammunition. A peculiar feature of this affair was that the lovers appeared to continue to make appointments by letter, even though they were sharing an apartment. Maybe in the Paris of 1910 that was the only way to communicate rapidly. In any event it was a mistake. The letters they wrote to each other were saved in a drawer in their apartment. In the spring of 1911 someone, apparently hired by Madame Langevin, broke into the apartment and stole the letters, placing them in Madame Langevin's hand. This was in effect the end of the affair, although it took a bit of time for the drama to play itself out. The most damaging letter was one that Marie wrote to Langevin from her sea-coast home in Brittany in September 1910. It was a very long letter in which Marie presented Langevin with what amounted to a step-by-step blueprint for extracting himself from his marriage and making his life with her. Here are some of the things she writes:

> There are deep affinities between us which only need a favorable life situation to develop. We had some presentiment of it in the past, but it didn't come into full consciousness until we found ourselves face to face, me in mourning for the beautiful life that I had made for myself and which collapsed in such a disaster,

you with your feeling that, in spite of your good will and your efforts, you had completely missed out on this family life which you had wished to be so rich in abundant joy . . . .

Then she has a few choice things to say about Langevin's wife:

> Your wife is incapable of remaining tranquil and allowing you your freedom; she will try always to exercise a constraint over you for all sorts of reasons: material interests, desire to distract herself and even simple idleness. . . .

Keep in mind that Madame Langevin was at this time taking care of a child that had been born a year earlier as well as three other children. Marie has something to say about this as well:

> If the separation took place, your wife would very quickly stop paying attention to her children, whom she is incapable of guiding and who bore her, and you could take up little by little the preponderant direction.

Marie also has detailed advice on how to proceed:

> It is certain that your wife will not readily accept a separation, because she has no interest in it; she has always lived by exploiting you and will not find that situation advantageous. What's more, it is in her character to stay, when she thinks you would like her to go. It is therefore necessary for you to do all that you can, methodically, to make her life insupportable. . .the first time she proposes that she could allow you to separate while keeping the children, *you must accept without hesitation* [The italics are in the original.] to cut short the blackmail she will attempt on this subject. It's enough now that Jean [Langevin's oldest child, who was then eleven.] continues to board at the Lycée and that you live in Paris at the school; you could go to see your other children at

Fontenay or have them brought to the Perrins' [This is a reference to the physicist Jean Perrin and his wife. Langevin had done some of his earliest research with Perrin.]; the change wouldn't be so big as you think and *it certainly would be better for everyone.* [Including Madame Langevin?] We could maintain the same precautions we do now for seeing each other until the situation becomes stable. . . .

This is not exactly the letter one wants to fall into the hands of one's rival, or the press. But both things happened. Before the letters become public there were rumors about the affair. As usually happens in these situations, people chose up sides. Einstein weighed in with the remark that Madame Curie "was not attractive enough to become dangerous for anyone,"* which shows that when one makes comments outside one's specialty, one runs the risk making a total fool of oneself. Meanwhile, Langevin moved back in with his wife, at least for awhile. Perhaps things would have calmed down except that both Langevin and Madame Curie were invited to the Solvay Congress in Brussels in 1911. This was too much for Madame Langevin, and she decided to make the letters public. Enter one of those figures that catalyze a situation and then vanish into relative oblivion.

In this case it was a newspaperman named Gustave Téry. Téry was someone who had been on most sides of most issues—all with equal vigor. In his younger years he had lampooned the Catholic Church and had taken Dreyfus's part in the famous affair. But by 1909 he had founded a newspaper named *l'Oeuvre* and had taken a sharp turn to the right. He was now a chauvinist and an anti-Semite who referred in articles to the "German-Jewish Sorbonne." One of the strange

*Quinn, *Marie Curie: A Life,* 313.

aspects of Madame Curie's scandal is that it took on anti-Semitic over-tones. None of the principals were remotely Jewish, but discontented anti-Semites like Téry managed find some way to blame the Jews for it anyway. Téry published a substantial portion of the letters, including the long one from Marie outlining her game plan. But he was not content with this. He commented incessantly in his newspaper, noting, for example, that Langevin was being referred to as the *"chopin de la Polonaise."* "Chopin" is an outdated French slang word meaning "patsy." The wordplay is so good that, considering the general low level of the rest of his writing, one wonders if he thought it up himself.

With the publication of the letters, the scandal could not be ignored. To add to everything, the Swedish Academy had just voted to confer on Madame Curie a second Nobel Prize. The day before the letters were published, Madame Curie wrote to Svante Arrhenius, a member of the Swedish Academy and one of her strongest supporters. She inquired whether, in view of all the rumors circulating about her personal life, it would be better if she stayed away from the ceremony. Arrhenius assured her, by letter, that all would be well and that she could come. But six days later he changed his mind. In the meantime, the letters had been published and Langevin had challenged Téry to a duel. Téry had named him explicitly in his newspaper, calling him a "boor and a coward"—dueling words. Pistols were chosen. On the morning of November 26th the two men met, armed and with their seconds. What happened next shows, at least to me, that Téry was a better and more interesting man than one would have thought. After the two men had lined up, Téry pointed his pistol to the ground, indicating that he had no intention of firing. Langevin then did the same, and the matter was settled when the seconds took the pistols and fired them into the air. Later, in his newspaper, Téry wrote ". . . however grave may be the errors made by Langevin in his domestic life, I obviously had scruples about depriving French science of a precious brain. . . ." One wonders what happened to Téry afterwards.

Arrhenius then wrote to Madame Curie informing her that if the Academy had known the facts outlined in her now published letters, they would not have awarded her the Prize. To her credit, she stood her ground replying to Arrhenius that "In fact the Prize has been awarded for the discovery of Radium and Polonium. I believe that there is no connection between my scientific work and the facts of private life. . . . I cannot accept the idea in principle that the appreciation of the value of scientific work should be influenced by libel and slander concerning private life." In any event, she did attend the award ceremonies in Stockholm.

One of the consequences of Langevin's duel, and the publicity surrounding it, was that he and his wife arranged a legal separation with her getting custody of the children. Marie Curie wanted him to seek a divorce, but Langevin refused, saying he would not take sides publicly against the mother of his children. This, it seems, ended their affair. Whether it also ended Marie Curie's *vie sentimentale* in general, I do not know, but I would guess yes.

It took many years before Madame Curie's reputation was restored in France. It might have taken longer except for the First World War, which gave the French something more serious to think about. Madame Curie devoted all her energies to service during the war. The most important thing she did was to create a portable X-ray diagnostic facility in an automobile she had managed to scavenge. By the time the war ended, she had created a fleet of 18 vehicles to be used as portable X-ray laboratories, and in which tens of thousands of wounded men were examined. Madame Curie not only organized all of this, but she personally trained the X-ray technicians and diagnosticians, and when necessary drove to these posts to help repair the machines. In 1916 she obtained a driver's licence for this purpose. At the age of 18, her daughter Irène also began to teach radiology. This began the collaboration between the two women that continued in Madame Curie's laboratory for the rest of her life.

In the years after the war, Marie Curie became a kind of icon— a world-renowned, instantly recognizable scientific figure of which there have been very few. Einstein and Stephen Hawking are examples. Madame Curie was quite capable of using her fame to advance causes she was committed to. One of them was to acquire an additional gram of radium for her laboratory. To this end, she lent herself to a successful fund-raising tour in the United States in 1921. Just before she returned to France, President Warren Harding presented her with a green leather case containing an hour glass that symbolized the actual gram of radium that had safely been stored elsewhere.

Madame Curie had always had a proprietary feeling about radium. In 1910, Rutherford proposed that she establish the universal standard for radiation activity based on the activity of radium. This is comparable to establishing the length standard of some known, but arbitrary, physical object. She did so by establishing the radioactivity of a gram of radium—a unit that was named the "curie" in honor of Pierre. This unit is still used, but in 1953 it was redefined to be the quantity of *any* radioactive material that produces 3.700×10^{10} decays a second. After Madame Curie established the standard she decided that she wanted to keep the actual sample in her laboratory "partly for sentimental reasons." It took some two years before the sample could be pried away from her and placed in the Bureau of Weights and Measures in Sèvres.

From the war on, Madame Curie's health was never really very robust. It is amazing to me that it took until 1934 for the effects of radiation to finally kill her. She died in a sanatorium in the French Alps on the 29th of June that year. She had gone through several cataract operations and suffered from lesions on her fingers from handling radium. The actual cause of death was pernicious anemia, surely brought about by radiation exposure. But for nearly a decade prior to her death, the most interesting work in her laboratory was being done by others—especially by her daughter Irène and her son-in-law Frédéric Joliot.

One of the most unsatisfying things in Ms. Quinn's book is her treatment of Joliot. It isn't what she says but what she leaves out— nearly everything. Joliot, who was born in 1900, was also a bit of an outsider. He came from a completely nonreligious family. After his father died, he had to leave the boarding school he had been attending for a public school. Then in 1920 he entered the École Municipale de Physique et de Chimie Industrielle, the very place Pierre Curie had taught, and where Langevin was now the director of studies. The entry in the *Dictionary of Scientific Biography* on Joliot notes that it was Langevin "who had a decisive influence on Joliot: he oriented the young man not only toward scientific research but also toward a pacifist and socially conscious humanism that eventually led him to socialism." In 1942 it led him to join the then-clandestine Communist Party. It probably also led him to his notable behavior in the Second World War, to which I will return in a moment.

Langevin, who seems to have maintained at least a friendship with Madame Curie, learned that she had a stipend to pay an assistant at her Institut du Radium and, in 1925, Joliot took the job. There he met Irène Curie who was also at the Institut and they were married in 1926. They had a daughter and a son. The daughter, Hélène, who was born in 1927, became a physicist, graduating first in her class from the very École Municipale of her grandfather. In 1949 she married the grandson of Langevin, who was also a physicist! I wish I knew more about their work, and whether they have children who are also scientists.

Most of the time that Joliot and his wife were at the Institut, they worked on separate projects. But there was a four-year period beginning in 1934, when they collaborated and produced the research for which they won the Nobel Prize in 1935.

This work was really in nuclear physics. It involved the production of radioactive elements by bombarding stable ones with nuclear projectiles. This is sometimes called "artificial radioactivity," but I think the name is misleading. It is really a kind of nuclear alchemy in which a stable nucleus absorbs some of the nuclear components of the bombarding particle and is transformed into a new isotope, which is radioactive. We now take this process for granted, forgetting that it had to be discovered by someone. In the course of this work they had some assistance from a young German physicist named Wolfgang Gentner. He was to reappear in their life in the Second World War. Joliot was one of the people who discovered that in nuclear fission, neutrons are released so that a chain reaction is possible. He made this discovery in 1939 and, unlike what our scientists did, published it in the open literature. It was read, for example, by the German nuclear physicists and was one of the reasons they then began their atomic bomb program. Joliot was fully aware of what his discovery meant. In fact, he ordered whatever heavy water was available in Norway and six tons of uranium oxide to try to make a nuclear reactor. When the Germans invaded Paris, he managed to spirit away this potentially dangerous material. Soon after the city's capture, he was visited by some German bomb scientists. They left Gentner as a kind of liaison to their program. But Gentner was a committed anti-Nazi. He made sure that nothing of value got to Germany and he shielded Joliot who had by now joined the Resistance. After the war, Joliot, although he was a Communist often in opposition to the government of General de Gaulle, was nonetheless largely responsible for creating the French nuclear energy program. He died in 1958, two years after Irène's death.

Eve Curie, Madame Curie's other daughter, later Mrs. Henry Labouisse, appears to have had an artistic rather than a scientific bent. I have not been able to find out much about either her or any children she might have had. However, under her maiden name, Madame Labouisse published one of those awful hagiographic biographies of her mother that I referred to in the beginning of this essay. Here is what she has to say about the "affair"—it is *all* that she has to say:

> Marie, who exercised a man's profession, had chosen her friends and confidants among men. And this exceptional creature exercised upon her intimates, upon one of them particularly, a profound influence. No more was needed. A scientist devoted to her work, whose life was dignified, reserved, and in recent years especially pitiable, was accused of breaking up homes and of dishonoring the name she bore with too much brilliance.
>
> It is not for me to judge those who gave the signal for the attack, or to say with what despair and often with what tragic clumsiness Marie floundered. Let us leave in peace these journalists who had the courage to insult a hunted woman, pestered by anonymous letters, publicly threatened with violence, with her life itself in danger. Some among these men came to ask her pardon later on, with words of repentance and with tears.

Fortunately, thanks to writers like Ms. Quinn we now know enough about Marie Curie to see her for what she was: a remarkable three-dimensional human being—not an icon.

Segrè

In 1926, Enrico Fermi received an appointment as a full professor of physics in Rome. He was just 25. He had already made several significant contributions to physics, the most important of which had to do with the quantum statistical mechanics of particles like electrons. It was the first of several discoveries for which Fermi deserved a Nobel Prize. (He was awarded the Prize in physics for 1938.) He had been studying in Germany and Holland and then obtained a temporary appointment in Florence. From there he was summoned to Rome. At the time, he was probably the only scientist in Italy who really understood modern physics. Until 1928, there was not even a text in Italian suitable for introducing graduate students to the subject. Fermi was determined to change all of that, and he began recruiting promising students who were not much younger than himself. One of the early recruits was Emilio Segrè, whose posthumous autobiography *A Mind Always in Motion* was published in 1994 by the University of California Press. Segrè died in 1989 at the age of 84. The autobiography ends when Segrè is 77 and was only published after his death because, as he writes "I tell the truth the way it was and not the way many of my colleagues wish it had been." Segrè was not a man with a great deal of diplomatic tact and his autobiography reflects this.

The route that led Segrè to Fermi is an interesting one. Segrè was born into a prosperous, nonobserving Jewish family that had lived in Italy for centuries. He conjectures that the Segrès had migrated to Italy when the Jews were expelled from Spain in 1492. His father owned a paper mill in Tivoli, but as a sort of unpaid public service he administered the nearby Villa d'Este, which was the property of the Archduke Francis Ferdinand (who never visited it). The elder Segrè took on the responsibility for maintaining the historic buildings. One of the most attractive parts of Segrè's book is his description of his childhood, which seems to have been rather idyllic. His fondest memories are of his father's brother Claudio, who took a special interest in his nephew. Claudio was an engineer and he encouraged Segrè's budding interest in science, especially technical gadgets involving electricity. By the time Segrè reached high school he was beginning to study more advanced physics on his own. For example he was reading books like James Clerk Maxwell's *Theory of Heat*. He describes how hard he found it and comments "I had not yet learned that in order to study physics, one has to use paper and pencil and work through the calculations as one goes along. Usually I read these books at school during boring classes that I disdained."

After graduating from high school, Segrè entered the University of Rome with the intention of becoming an engineer and perhaps working in his father's paper mill. He also discovered mountaineering and, while he does not dwell on his abilities, the climbs that he describes—such as what is known as the "Italian route" on the Matterhorn—were done without guides and were quite serious. It was one of his climbing friends, a fellow engineering student named Giovanni Enriques, who first called Segrè's attention to the arrival of Fermi. Segrè went to a lecture of Fermi's and was extremely impressed, but nonetheless continued his engineering studies which he was finding more and more distasteful. However, in the spring of 1927 Enriques introduced him to the then 25-year-old physicist Franco Rasetti, who was in Rome and a close friend of Fermi's. He was also a mountaineer, and during the course of some climbing expeditions Rasetti persuaded Segrè to meet Fermi with the notion that he might like to switch from engineering to physics. Fermi was also a mountain walker and he came along on a hiking expedition that included Segrè. During the course of it, Fermi quizzed Segrè to get a sense of whether he might be a suitable student. Keep in mind that Fermi was then 26 and Segrè 22. Despite the fact that they were near contemporaries, there was never the slightest question then, or afterwards, who was the teacher and who was the student.

Segrè made a favorable enough impression to become Fermi's first graduate student in 1928, an extraordinary opportunity. As a physicist Fermi was, at least in the modern age, unique. He worked both experimentally and theoretically in every branch of physics and may well be the last physicist who was able to master the entire field. It is just too complicated to do that now. His self-confidence was total. It is little wonder that when Fermi's group in Rome began giving themselves ecclesiastical nicknames, Fermi became known as the Pope;

Rasetti was known as the Cardinal Vicar. "I," Segrè writes, "[was known as] the Prefect of Libraries, because I was interested in the library; however, I was also the Basilisk, because I was supposed to spit fire when mad. . . . "

Fermi genuinely liked having students, if they were bright enough, and began tutoring Segrè and Rasetti privately. I can imagine what that was like. When I was a graduate student at Harvard in the early 1950s, Fermi visited Cambridge to give a series of public lectures. In addition, he gave an impromptu, semiprivate lecture for about a half-dozen of us. He chose to talk about a standard problem in quantum mechanics—scattering from a so-called square well for the aficionados—but which he had a novel way of looking at. After he finished, one of our bolder colleagues challenged the rigor of what he had just heard. Fermi then gave a second impromptu lecture on the same subject with more rigor. Each time the challenger interrupted, Fermi produced a new level of rigor. After a couple of these demonstrations, his interlocutor gave up. (I know several examples of Fermi's having asked what seemed to have been spontaneous questions in the course of someone's lecture that were so deep that they began entirely new fields of physics.) But I also got the impression from this performance and from Segrè—who knew Fermi very well and wrote an excellent biography *Enrico Fermi: Physicist* (University of Chicago Press, 1970)—that Fermi was a pretty cold fish. He did not seem to be personally accessible at all. I recall going up to him after his talk to bring greetings from an uncle of mine who had been Fermi's colleague and neighbor at Columbia and later at the University of Chicago. Fermi nodded his head and walked off, without saying a word.

After getting his degree, Segrè wanted to continue working with Fermi. Unlike most young researchers, Segrè was not strapped for money since his father had agreed to support him. In the meantime,

Segrè was called for military service, which he did, not too disagreeably, as a second lieutenant in the antiaircraft artillery stationed near Rome. The proximity enabled him to keep in touch with the laboratory. After his service, he rejoined Fermi's group. By 1930 he was publishing his own experimental work and was then able to spend a year as a traveling scientist in Holland and Germany. While in Germany, he appears to have fallen in love with a young German woman. He might, it seems, have married her, but as he got to know her better he realized that she was a German nationalist who was becoming more and more deeply committed to the Nazi movement. Eventually he broke off the relationship. Interestingly, members of Segrè's family, especially the older generation, had joined the Fascist Party in Italy, something that was not uncommon for upper-middle-class Jews to do at the time. But Segrè realized that the Nazi Party in Germany was something else. Above all, it was not for middle-class Jews.

To me, at least, none of the work that Segrè did during this period seems very interesting. The really interesting work started in 1933 when Fermi began his studies in nuclear physics. This was the work cited in his Nobel Prize. The neutron had been discovered a year earlier. It was an ideal nuclear probe since, being electrically neutral, it could penetrate deeply into the atomic nucleus without being repelled by the electrical force of the protons. Fermi began a series of experiments in which he bombarded one element after another with neutrons to see what would happen. The result was nuclear alchemy—the nuclei were transformed into radioactive isotopes, which could then decay into entirely different nuclei. Working up the periodic table, the group eventually came to uranium. They expected, and indeed thought they had witnessed, the transformation of uranium into its neighboring heavy nuclei. They had no idea—no one had any idea—that uranium could be made to fission (split) into two light nuclei such as barium and krypton when bombarded by neutrons.

In 1934 the group made a very odd discovery. They found that if they did their experiments on a wooden table, as opposed to a marble shelf, the silver they were irradiating at the time became much more active. Somehow the presence of the wood enhanced the nuclear reaction. Puzzled, Fermi devised a test. He put a filter between the neutron source and the target to see what would happen. His first choice as a filter was lead, but for reasons he could never explain, at the last minute he chose to use paraffin. Miraculously, the neutrons filtered by the paraffin produced extraordinarily enhanced rates for the nuclear reactions they were inducing. At this point Fermi went home for lunch and his habitual afternoon siesta. By the time he came back at three o'clock, he had pieced it together, creating a very important new branch of experimental nuclear physics.

What Fermi realized was that paraffin, in the language of the reactor physicists, acts as a "moderator." The neutrons striking the carbon and hydrogen nuclei in the paraffin bounce off, losing energy. They are slowed down, and after a few collisions move about with the same speeds as the molecules in the paraffin. They have been, as physicists say, "thermalised." What no one expected was that the slow neutrons would be more effective in collisions than the fast neutrons. As Fermi was later able to show, this result follows from the quantum theory of these collisions. But, having realized the efficacy of slow neutrons, it now became a matter of repeating the original experiments, this time using slow neutrons.

Early in 1935 they did the slow-neutron experiment with a uranium target. This is what Otto Hahn and Fritz Strassmann did in Germany in 1938, when they actually discovered fission. When I learned that Fermi and his group had done the same experiment in Rome three years earlier, I was completely baffled as to why they had not discovered fission then. I had a chance to ask Segrè about this a few years before his death during an evening I spent with him and several physicists from Columbia University, where he had given a lecture. His answer gave me the shudders. In order to shield their detectors from unwanted radiation, they had covered the uranium target with aluminum foil. This foil kept them from seeing the very energetic pulses coming from the uranium fission that was taking place. They had missed discovering fission because of the shielding effect of this aluminum foil.

When Segrè told me this anecdote I was stupefied. With Fermi's genius for understanding the significance of experimental results, it was clear that if he had seen unexpected energy pulses coming from uranium, he would have been led to the discovery of fission. There was even a chemist named Ida Noddack who suggested this possibility in a speculative article, a copy of which she sent to the Rome group at this time. They dismissed it, since it seemed inconsistent with some of the data on nuclear masses. Surely if they had seen energy pulses they would have reexamined all of this data. But that would have meant the race to build an atomic bomb might well have started in 1935 rather than 1939. If so, the Second World War could have been nuclear from the beginning, or perhaps there would have been no Second World War given the prospect of nuclear weapons. All of this because of some aluminum foil! When I suggested these hypothetical possibilities to Segrè he did not appear very interested. What happened, happened, and that was that. Indeed, while he does describe the situation with the aluminum foil in his biography of Fermi, he does not seem to find it significant enough to include in his autobiography. Historical speculation appears not to have been Segrè's cup of tea.

The Italian university system then (and I think still today) required that aspiring university professors begin their careers in provincial universities. Some academic roads eventually led to Rome, but slowly, unless one was a Fermi. Segrè began his professorial career in 1936 at the University of Palermo in Sicily. By this time he had married a German-Jewish woman named Elfriede Spiro, whom he met in 1934, a year after her family had been forced to emigrate to Italy. Palermo was something of a backwater in physics and Segrè set about to change that. He had already made two visits to the United States with Fermi and in the summer of 1936 he returned again, this time to visit the University of California at Berkeley, where Ernest O. Lawrence had constructed the first cyclotron—a machine that he had invented.

Much of Segrè's autobiography revolves around his complex and often adversarial relationships with other physicists, including people like Lawrence, Oppenheimer and the chemist Glen Seaborg. This is where Segrè's lack of tact shows. His relationship with Lawrence began amiably enough. He was given permission to take back to Palermo some radioactive detritus that had been manufactured haphazardly in the cyclotron. Upon Segrè's return to Palermo, Lawrence sent him another batch, which happened to contain a molybdenum foil that had been irradiated in the cyclotron. Using this foil, Segrè made the one important discovery that, as far as I can see, was basically his own. (He did have the collaboration of a chemist named Carlo Perrier, but it was really Segrè's experiment.) His other important experiments, including the one in which the antiproton was discovered and led to Segrè's Nobel Prize, were done in collaborations in which Segrè's contribution was not entirely clear. I will come back to this when I discuss Segrè's Nobel Prize.

In any event, using the molybdenum foils, Segrè and Perrier were able to identify a new element, one of the radioactive decay products of the molybdenum. The new element was eventually named "technetium." It turned out to have very important applications in nuclear medicine. One may well wonder why Lawrence himself did not discover this element, which was present in his own molybdenum foils. Segrè delivers a somewhat harsh, but I think correct, judgment of Lawrence. Lawrence, he notes, was not terribly interested in science per se. He was probably not even a very good scientist. He *was* interested in building bigger and bigger cyclotrons. In this, Segrè acknowledges, he was a genius. All of us who work in the field of elementary particles can be grateful to him for what has been discovered using his machines and their descendants.

In 1938, Segrè was back in Berkeley for what he thought was a summer visit. But as events played themselves out, he spent the next 35 years either at Berkeley or, during the war, at Los Alamos. In 1938 Mussolini had promoted a *Manifesto della razza*—*race* manifesto—which banned Jews from holding jobs in universities. Segrè's job in Palermo disappeared. Fortunately, he was not entirely unprepared. Family money had been taken abroad for just such a contingency. In his book Segrè writes a great deal about his family's business affairs—perhaps too much—but we never really know how much money was involved. Apparently enough so that, a few years later, he was able to use family money to buy a house in Berkeley. In any event, it was during the summer of 1938 that Segrè met people like Oppenheimer and the chemist Glen Seaborg who were to play an essential part in his life for the next several years. What is clear from his account of these two men, is that he didn't like either of them. I find Segrè's characterization of Oppenheimer understandable, but off the mark. He writes:

> Oppenheimer and his group did not inspire in me the awe
> that they perhaps expected. I had the impression that their cel-
> ebrated general culture was not superior to that expected in a boy

who had attended a good European high school. I was already acquainted with most of their cultural discoveries, and I found Oppenheimer's ostentation slightly ridiculous. In physics I was used to Fermi, who had a quite different solidity, coupled with a simplicity that contrasted with Oppenheimer's erudite complexities. Probably I did not sufficiently conceal my lack of supine admiration for Oppenheimer, and I found him unfriendly, even if covertly for a good part of my career, except when he wanted me to join his team at Los Alamos.

Unfortunately, Segrè does not explain what he means by the phrase "unfriendly, even if covertly" so we are left to speculate on what this innuendo signifies. Segrè's characterization of Oppenheimer's "erudite complexities" is right, as far as it goes. I will never forget an occasion when I was a very junior visitor at the Institute for Advanced Study in Princeton in the late 1950s, when Oppenheimer was its director. He used to call us into his office for little "confessionals" concerning our most recent activity. On this occasion, my only recent "activity" had been reading Proust, whom I had just discovered. Since I had already decided that I had no future in physics anyway, there was no point in making something up; I told Oppenheimer the simple truth. He looked at me kindly and said that when he was my age he had also read Proust—but in French, and by flashlight, while touring Corsica by bicycle. If that was not "erudite complexity," I cannot imagine what would be.

But this characterization alone could not possibly explain Oppenheimer's role in American physics. I think one could make the case that Oppenheimer and a tiny handful of other Americans, like I.I. Rabi and Linus Pauling, were responsible for bringing modern physics to this country in the late 1920s and early 1930s. Oppenheimer was one of the greatest teachers of physics who ever lived, and he was also a first-rate physicist. Among many other things, as we have seen, it was Oppenheimer and his students who in the late 1930s realized the possibility of gravitational black holes. It should also be understood that Oppenheimer's productive years were truncated by the war. He threw himself body and soul into the work on the atomic bomb. To get some idea of the complexity involved, a reader with a little technical background might want to read a book entitled *Critical Assembly* (Cambridge University Press), a collection of fascinating related essays on the technical history of Los Alamos from 1943 to 1945. The things that Oppenheimer had to understand—and make decisions about—are mind boggling. One wonders if anyone else would have had the intellectual capacity to do it, never mind the personal magnetism needed to keep such a vast collection of scientific prima donnas functioning including, one may add, Segrè himself.

I imagine that Oppenheimer did not find Segrè's work intellectually very interesting, and since he was a person who had even less tact than Segrè, he very likely made that clear one way or another. But, in truth, I do not find Segrè's work intellectually very interesting either. It is, of course, admirable to have discovered a new element. But how can one compare this with the discovery of the quantum theory, which contains within its riches the explanation of why there is a periodic table of elements at all? The creators of quantum theory, like Bohr, Dirac and Pauli, were Oppenheimer's contemporaries and colleagues. These were the people Oppenheimer really respected. If there was a tragic component to Oppenheimer's life, it was that he knew he had abilities on a par with these men, but he was never able to do scientific work at quite their level. Oppenheimer was well aware of this and sometimes brought it up himself. He was a perfect example of the fact that extraordinary intelligence is not enough to do great scientific work. Something else is needed. "Solidity," Segrè's word, may not be a bad description of what was missing in Oppenheimer; men like Fermi and Bohr had it in abundance.

To be fair, it must be said that Segrè knew that, despite his eventual Nobel Prize, he would occupy a fairly modest place among 20th-century physicists. He recalls a conversation he once had with Fermi:

> "Emilio you could take all your work and exchange it for one paper of Dirac's and you would gain substantially in the trade," Fermi once said to me. I knew this to be true, of course, but I answered: "I agree, but you could likewise trade yours for one of Einstein's and come out ahead." After a short pause, Fermi assented. I know of scientists who cannot resign themselves to being inferior to contemporaries, with dire consequences for their personalities and happiness.

I have no personal knowledge of Segrè's relationships with Lawrence and Seaborg, other than what he writes in his book. His account is not implausible and it is not very attractive. By the time these relationships developed, Segrè had managed to bring his wife and child to California from Italy; their parents and a brother remained. The brother hid in the hills behind Tivoli during the war while his father was hidden in the papal palace, protected by a high-ranking prelate, Monsignore Carinci. His mother, tragically, was caught by the Germans and murdered. When it became clear to Lawrence that Segrè could not return to Italy and that he was no longer a visitor with another job to go back to, he immediately cut Segrè's salary from $300 a month to $116. Segrè was "stunned" but later came to realize that in his own brute-force way Lawrence was being fair. His salary as a visitor had been anomalously high, and now—like it or not—he would earn something closer to the going rate. Segrè was able to supplement this with his private funds.

Segrè's relationship with Seaborg is another matter. Seaborg was seven years younger than Segrè. The two men began collaborating on technetium isotopes as soon as Segrè arrived at Berkeley. But Segrè's view of Seaborg soon darkened. In his book he writes:

> What was unusual in Seaborg was the long-term planning he diligently applied to everything. In 1941 he would say: in 1946 I shall be dean; in 1948, chancellor of the University of California [It actually took him until 1958.]; in 1955, senator for California [He never was.], and so on, and he never lost sight of his aims. In 1938 he always dressed in a blue suit, with a tie, differently from his colleagues, because he thought that these clothes would help him become a full professor, a small first step in the grand design. Ultimately, he devoted much effort to public service as chairman of the Atomic Energy Commission and many other organizations, receiving more than fifty honorary degrees and collecting pictures of himself with a number of presidents of the United States and other such figures.

While this portrait is not very flattering, it is only a prelude to what follows. Segrè makes it clear that he holds Seaborg, and even Seaborg's wife, responsible for his not getting his fair share of credit for the work done at Berkeley on transuranic elements like plutonium; only Seaborg and Edwin McMillan shared the 1951 Nobel Prize in chemistry for that. Seaborg's wife—née Helen Griggs—had been Lawrence's secretary. Segrè came to believe that she gave her husband some papers from Lawrence's files containing data on plutonium that Segrè and his collaborator Joseph Kennedy had collected. A joint paper written with Seaborg was held up for publication during the war, which also clouded the issue of credit for this work. There was a probable Nobel Prize at stake. Segrè writes "After the war I had started thinking that my work on the new chemical elements and on radio-chemistry might bring me that distinction. I saw Seaborg's efforts at getting it on similar grounds, but I did not know how to stake my claim. I hoped that the Nobel committee would somehow split the award." The committee chose not to, and when Segrè heard the news, he was, "deeply disappointed," which I imagine is a considerable understatement.

I next want to discuss Segrè's own Nobel Prize and the ill feelings that it led to. But first let me say a little about the Prize itself. In physics, and in the other scientific disciplines, it can be shared by no more than three people. It cannot be given posthumously. Finally, the award is given by the Swedish Royal Academy which is, after all, a *national* academy of science like many others. What makes this prize special, apart from the large sum of money—which is now no longer unique for scientific prizes, nor that large considering it is usually shared and, at least in the United States, it is taxable—is the list of people who have won it. People like Einstein confer honor to the Prize, rather than the other way around. Each year, the Royal Academy sends out nominating ballots to a large number of physicists. One year I got one, but I have now forgotten whom I nominated. Clearly, some nominations carry much more weight than others. A nomination by, say, Einstein obviously carries real weight as opposed, say, to one by me. The members of the Academy even have consultations with such individuals. I know personally of an example. It was told to me by the person involved—a previous Nobel Laureate. He told me that he spent an afternoon in a physics library with a senior member of the Academy going over some published physics papers in order to clarify priority and credit for work that was being considered. Indeed, it was awarded a Nobel Prize that year. As a rule, the deliberations and the nominations are kept secret for decades. It is very interesting to read some of the nominating letters years after the fact.

People who understand the system and are sufficiently motivated can try to take matters into their own hands and this is what Segrè did. It is this part of Segrè's book I find the most unappealing. He seemed almost desperate to win the Prize and when he finally did, it was, I will argue, somewhat tainted.

He opened his campaign in 1954 on a visit to Brazil. There he met an old friend, George de Hevesy, who had won the Prize for chemistry in 1943 for his work on using radioactive materials as biological tracers. Segrè knew that Hevesy had the ear of the Nobel committee. He wrote "We [he and Hevesy] were friends and I could speak freely to him. Thanks to his Swedish connections he knew many of the secrets of the Nobel Committee, and he told me that I had not been specifically nominated in the year 1951, which had automatically eliminated me. He advised me to try to interest Fermi. I did not do so because I knew perfectly well that Fermi could not be influenced in matters such as competitions and awards." It doesn't seem to occur to Segrè that even if Fermi had been approachable, there was something distasteful about lobbying him, or indeed anyone, for the award. If it is to mean anything, it should be given for the greatest contribution, not the greatest ability to hustle for votes. How then did Segrè finally win the Nobel Prize?

In 1931, the great British theoretical physicist P.A.M. Dirac suggested the existence of antimatter—specifically the existence of the antiparticle to the electron, now known as the positron since it has a positive electric charge that exactly balances the electron's negative charge. In the same year, C.D. Anderson found the positron in cosmic rays. Over the next decades the theory of antimatter was clarified. It led to the prediction that for every particle there is an antiparticle with definite properties. The particle and its antiparticle must have exactly the same mass, they must have equal and opposite charges and, if they are unstable, they must decay at the same rate. When a particle and an antiparticle meet they can annihilate each other, creating an assortment of energetic reaction products. In this spirit, it was assumed that the proton, and also the neutron, would have antiparticle counterparts. Specifically, the antiproton would be an essentially stable object with a negative charge and the same mass as the proton. It could bind with antineutrons and create a universe of antimatter. Physicists tend to express these particle masses in terms of energy units since Einstein's relation $E=mc^2$ means that mass and energy are essentially the same thing. The energy unit one uses here is the electron volt. In these units the electron has a mass-energy of about a half-million electron volts, while the proton, the neutron and their antiparticles have mass-energies of about a billion electron volts. Therefore to create an antiproton requires an energy in the billions of electron volts.

Prior to the war, cyclotrons could produce proton energies in the thousand-electron volt range. After the war, new machines with even larger magnets raised the range to millions of electron volts. But in 1954, using some new technological ideas, a machine was constructed at Berkeley that could produce proton energies of 6.2 billion electron volts—6.2 BeV. It was appropriately named the "Bevatron." The energy of 6.2 billion volts was not chosen haphazardly. This, it turns out, is about the minimum energy needed to produce an antiproton in a collision between two protons. To balance the various electric charges and other conserved quantities, the simplest reaction that works is proton colliding with a proton, producing three protons and an antiproton—symbolically $p + p \rightarrow p + p + p + \bar{p}$. The machine had been built to study just this reaction. When it was produced, some wag suggested that the *machine* should be given the Nobel Prize. So far the Nobel Committee seems not to have considered awarding prizes to machines.

It is not surprising that several researchers at Berkeley had the idea of doing the antiproton experiment. I imagine there was a great deal of infighting and politicking over which of the groups would be awarded the necessary time on the machine. Segrè does not say much about this in his book except to remark that "Several Berkeley groups started the hunt. My group had for some time studied the problem and prepared for it." For purposes of this experiment Segrè's group consisted of three other physicists besides himself. There were Owen Chamberlain and Clyde Wiegand who had been students at Berkeley. They had joined Segrè at Los Alamos where they did very important work on the properties of plutonium. Finally, there was Tom Ypsilantis, who was a decade younger than the others but was already an experimenter with great promise. In seniority, Wiegand, who was some five years older than Chamberlain, was next in age to Segrè. Wiegand was acknowledged to be an electronics genius whose wizardry was absolutely essential in making the experiment work. It was not easy. Ultimately, they detected about one antiproton per 100,000 events involving other particles, which meant they saw only a few good events per hour.

The paper announcing the discovery was published in the *Physical Review*—our physics trade journal—in 1955 under the title "Observation of Antiprotons." It was signed in alphabetical order O. Chamberlain, E. Segrè, C. Wiegand and T. Ypsilantis. It was a logical candidate for a Nobel Prize, but to whom? Segrè writes, "Needless to say, before the announcement I did not know if and how a prize given for the antiproton would be divided between Chamberlain, myself, Wiegand and Ypsilantis, since the paper reporting the discovery had been signed by all four of us in alphabetical order." If Segrè had no clear way of making this decision, then how could anyone else? But that is exactly what the Nobel Committee did. They awarded the 1959 Prize in physics to Segrè and Chamberlain for the discovery of the antiproton, leaving the others out. Since the deliberations are secret, only the Committee knows why.

Segrè's reaction to this curious award, at least as expressed in his book, seems to show an almost complete lack of understanding of the feelings of those involved. This is especially striking since *he* thought he had been unfairly treated in the work that led to the Prize awarded to Seaborg and McMillan. Here is some of what he writes. There isn't much:

> After the discovery of the antiproton and connected pub-
> licity, the moods of Owen and of Clyde separately darkened. Owen
> wanted to be more independent than he already was, which was
> hardly possible. [Why it was "hardly possible" is not explained.]
> He wanted to have his own group, but our group was so small
> that I felt that further splitting would impair its efficiency. Owen
> was then invited to go to Harvard. . . . On his return, he started a
> small separate group. Clyde too, wanted to go it alone, and above
> all to work independently of me and Owen. Perhaps he wanted
> to show his personal prowess, although his ability was widely
> recognized, above all by me and his other colleagues in the group.
> It is possible that even Ypsilantis had similar wishes, but being
> younger, at the beginning of his career, and of a sunny disposi-
> tion, he was less affected.

It never appears to dawn on Segrè that there might be something more at stake here than the fact that Wiegand's ability was recognized by *him*. He does not show the slightest empathy for how Wiegand, a senior physicist on the experiment, must have felt having been left out of the Prize, although he does seem to realize that something was wrong. He describes the contributions of each one and adds, "Nor did I think that my contribution was as negligible as it perhaps then appeared to Owen and Clyde." What *was* Segrè's contribution to this experiment? He never makes this clear in his book. Segrè writes about it in an oddly detached way, unlike the way he writes about the experiments he took part in during the earlier part of his career. In those passages one can share some of his feelings as he actually comes to grips with a scientific discovery. He writes about the antiproton experiment as if it had been done by somebody else. Nowhere does he give us a feeling for what it was like to actually take data and make this discovery.

Wiegand's feelings, and Segrè's lack of understanding of them, seem to have persisted for many years. In 1985, four years before Segrè's death, there was a symposium in Berkeley celebrating the 30th anniversary of the experiment. Segrè describes the occasion in a footnote. After the opening speeches, he writes, "Clyde Wiegand followed for about thirty minutes. He gave a detailed account of the work performed by himself and Chamberlain. Among other things, he said that he and Chamberlain had started planning the experiment secretly, outside of regular working hours. He mentioned Ypsilantis only peripherally, in connection with the addition of a counter to the apparatus. He never mentioned me, as though I had not existed. I was saddened by this performance. It must represent Wiegand's present state of mind; this must be his recollection of the experiment." That is all he says. He never mentions that Wiegand's recollection is wrong, nor does he tell us what he himself actually did; only that he was saddened by the "performance."

I know of only one example—there may be others, but I don't know of them—when a scientist who won a Nobel Prize apologized, and in writing, to a collaborator who didn't. The occasion is described in the posthumously published autobiography, *My Life,* written by the very distinguished German-Jewish theoretical physicist Max Born. In the mid-1920s, when Born was still in Göttingen before being forced out by the Nazis, Werner Heisenberg came to him with a manuscript of a paper he had just written about atomic mechanics. Born studied it and realized that what Heisenberg had unwittingly done, was to use a branch of algebra involving matrices—arrays of numbers that obey certain algebraic rules. Born and a student, Pasqual Jordan, then created something that is now known as "matrix mechanics"—a version of the quantum theory. Soon afterwards, they were joined by Heisenberg, and the three of them made several very important applications of the new mechanics. In 1932, Heisenberg was awarded the Nobel Prize in Physics for this work, and Born and Jordan were left out. Heisenberg felt so badly about this that he made a special trip to Switzerland so he could mail an uncensored letter to Born, who was by then living in England. This is what he wrote:

> Dear Mr. Born:
> If I have not written to you for such a long time, and have not thanked you for your congratulations, it was partly because of my rather bad conscience with respect to you. The fact that I am to receive the Nobel Prize alone, for work done in Göttingen in collaboration—you, Jordan and I—this fact depresses me and I hardly know what to write to you. I am, of course, glad that our com-

mon efforts are now appreciated, and I enjoy the recollection of the beautiful time of collaboration. I also believe that all good physicists know how great was your and Jordan's contribution to the structure of quantum mechanics—and this remains unchanged by a wrong decision from outside. Yet I myself can do nothing but thank you again for all the fine collaboration and feel a little ashamed.

With kind regards,

Yours,

W. Heisenberg

I have found much in Heisenberg's life to reproach, but in writing this letter to Born he shows his best nature. Would that Segrè, in his last scientific testament, had found it within himself to do the same.

This essay, in a slightly different form, was published in March 1994 in the New York Review of Books. *Sometime later Segrè's widow Rosa— his first wife had died some years earlier—wrote an impassioned letter to the* Review *complaining about the critical character of my essay. I wrote a short reply in which I noted that anyone who begins a book—of which, incidentally, Mrs. Segrè appears to have collaborated—with the statement that "I tell the truth the way it was and not the way many of my colleagues wish it had been," should not be surprised at the reactions it provokes.*

Linus Pauling

During a talk in New York City, I mentioned how much pleasure I took in reading about the discoveries made by scientists in their various investigations of the nature of the world, and stated that I hoped I could live another twenty-five years in order to continue to have this pleasure. [The talk in question was given in 1966. This excerpt, which is from an article entitled "My Love Affair with Vitamin C" was written in 1992. Pauling died in August 1994 at the age of ninety-three.] On my return to California I received a letter from a biochemist, Irwin Stone, who had been at the talk. He wrote that he was sending me copies of some papers he had just published, with the general title "Hypoascorbemia: A Genetic Disease," and that if I followed his recommendation of taking 3,000 milligrams [3 grams] of vitamin C [daily] . . . , I would live not only twenty-five years longer, but probably more. [Pauling did.]*

Sometime in the 1970s I began wondering whether Linus Pauling, winner of two Nobel Prizes—chemistry in 1954, peace in 1962—had become a crank. In 1970 he had published a best-selling monograph entitled *Vitamin C and the Common Cold*. His thesis was that daily megadoses—several grams—of vitamin C could prevent or help cure many diseases—the common cold being the prime example. But Pauling was not content simply to publish his views, which were seen as unsound by many authorities. One could often find him on television with his high-pitched voice, his aureole of white hair and his faintly rictal grin promoting the virtues of vitamin C like a carnival shill. He was also giving interviews to publications like the *National Inquirer* and *Midnight* and was suing various other publications that disagreed with him. In short, he appeared to many people to have become unhinged.

It was about this time that I began an odd "correspondence" with Pauling. I put "correspondence" in quotation marks because it involved his sending me reprints of his papers and monographs along with requests for commentaries. I invariably responded that I had none, since I did not know enough about the subject. That was the simple truth. This lasted almost until his death when I received yet another packet of papers. These dealt with his contention that a recently discovered state of matter—so-called "quasi-crystals"—was based on a misreading of the data. Here, at least, the discussion was about physics, which enabled me to ask the opinion of colleagues. They assured me that, in this instance, Pauling was simply wrong.

*Unless otherwise noted, all the quotations in this chapter are from the following collection: Barbara Marinacci, ed., *Linus Pauling in His Own Words* (New York: Simon & Schuster, 1995).

I found Pauling's behavior extremely puzzling. Why was a man who had won two Nobel Prizes and was considered one of the most productive scientists of his generation behaving this way? I had never met Pauling, nor had I ever read a serious biographical study of him. So I was at a loss. But, I now feel that, in a sense, I have "met" Pauling. Our "meeting" took place through a series of new biographical studies of him, especially *Force of Nature* * by the Oregon-based journalist Thomas Hager, who interviewed him extensively over several years. After reading Hager's book and others, and by talking to a few people who knew him, I now feel that I have a better idea of what made Pauling tick. However, I have concluded that it would take a novelist, rather than a biographer, to put all the pieces together.

> When I was eleven, with no outside inspiration—just library books—I started collecting insects. Not only did I collect insects, I also read about insects. . . . At the time, I was interested only in insects! Which is why, before I got interested in chemistry itself, I began to need chemicals.

The career of Linus Pauling, rising from a childhood of emotional and economic deprivation to become a great scientist, is a story that could have been taken directly from an American romantic novel. I emphasize "American" because, from the time of the Second World War almost to his death, Pauling was accused—absurdly—of being "un-American." He was persecuted by the FBI. His passport was confiscated and, in 1951, the House Un-American Activities Committee named him one of the foremost Americans involved in a "campaign to disarm and defeat the United States." It is difficult to imagine anyone

---

*Thomas Hager, *Force of Nature* (New York: Simon & Schuster, 1995).

who was more "American" than Pauling. His father, Herman Henry William Pauling, son of German immigrants, had come by stagecoach to Condon, Oregon, in 1899. A year later, he married "Belle," the beautiful daughter of the town's founding father, Linus Wilson Darling. Darling's wife, Alcy Delilah Neal, could trace her roots to the Revolutionary War.

Pauling's father was a pharmacist, but after the pharmacy he tended was sold, he moved to Portland to find work. Pauling was born there on February 28, 1901. From all accounts, Pauling and his father had a special understanding. His father appears to have realized that his young son had unusual intellectual gifts, which he tried to nurture. But all of this came crashing down in June 1910 when Pauling was nine—his father suddenly died. By this time, there were three children—Linus and two younger sisters—and his mother simply could not cope, financially or emotionally. Pauling seems to have had two reactions to this: First, he simply blocked out any overt emotional response to the loss of his father. Second, and more strikingly, he treated his distraught and demanding mother as irrelevant. Indeed, throughout his life, Pauling had a difficult time dealing with his emotions. One can understand why. One can also understand why he seemed so stubbornly single-minded about whatever he focused on. Pauling learned to shut out anything that threatened to upset his emotional equilibrium.

What Pauling focused on first and foremost was science. It became a refuge from the emotional chaos that surrounded him. At first it was insects, collecting and reading about them. Then a former business acquaintance of his father supplied him with chemicals like potassium cyanide(!) for killing and preserving them. Next he turned to minerals. At age 12, he made a lifelong friend of a boy named Lloyd Jeffress who had a homemade chemistry set in his basement. From the first time Pauling saw Jeffress make two substances turn into a third in a chemical reaction, he decided that this was going to be his life's work. He also realized that being a chemist meant going to college, for which there was no money and no family tradition. His mother was vehemently opposed. She wanted him to take a job and help support her and her two other children. Pauling simply ignored her. By working at all sorts of menial jobs and living on next to nothing, he managed to begin attending the Oregon Agricultural College—now Oregon State University—in the fall of 1917. It was the only college he could afford.

Fortunately for Pauling, the school had just expanded its chemistry program. Before long, the most knowledgeable instructor in the program was Pauling himself. At age 18, his junior year, Pauling was offered—and accepted—a job as a part-time instructor in courses he had taken the previous year. By this time, he was reading more in the contemporary chemistry journal literature than anyone else on the faculty. Not only did teaching enable Pauling to earn his way through college in a more agreeable way than chopping wood, but it enabled him to meet his future wife, Ava Helen Miller. Miller was a freshman taking Pauling's chemistry class when he was a senior. In addition to being beautiful she was, Pauling later recalled, "in some ways more intelligent than I—as a test we both took, early in our marriage, proved her to be. Not only was she quicker, but she had more correct answers." The two of them fell in love and wanted to get married, but as both sets of parents objected the marriage was put off for a year until 1923. Meanwhile, Pauling had gone off to graduate school—something his mother also opposed. Pauling again ignored her, although he did borrow a thousand dollars from an uncle to give his mother so he would not have to support her.

Pauling applied to several graduate schools, informing all of them that he would have to work part-time in order to pay his way. He was accepted at Harvard but was told that, if he took their offer of a half-time instructorship, it would take him six years to get his Ph.D. He decided that this was too long, although, by today's standards it does not seem excessive. That left Berkeley and the California Institute of Technology—Cal Tech—which had recently changed its name from the Throop Institute. Each had a great chemist on its faculty: G.N. Lewis at Berkeley and A.A. Noyes at Cal Tech. Many years later Pauling explained why he didn't go to Berkeley. "I heard a story—probably it's apocryphal—that when Lewis looked over the several dozen graduate applications to the Berkeley chemistry department in early 1922, he came to one, looked at it, and said, 'Linus Pauling, Oregon Agricultural College. I never heard of that place.' So my application went into the discard pile." In any event, before he heard from Berkeley, he was offered a fellowship at Cal Tech that would pay his tuition plus $350 a month for working as a teaching assistant. He accepted and remained there until he resigned, under rather unhappy circumstances, in 1964.

Cal Tech turned out to be a marvelous choice. He was rapidly able to fill the gaps in his knowledge of mathematics and physics. Moreover, Noyes had brought to the United States from Europe the then rather new experimental subject of X-ray crystallography—the use of X-rays to study the structure of crystals. Noyes assigned Pauling to a newly created X-ray laboratory run by a young professor named Roscoe Dickinson. The two of them used X-rays to determine the structure of molybdenite. This was Pauling's first scientific discovery—the first of hundreds. He must have made a considerable impression on Noyes since Noyes began one of those Machiavellian campaigns (the sustenance of academic life) to keep Pauling out of the clutches of G.N. Lewis at Berkeley. One of the strings he pulled was to arrange for a Guggenheim Fellowship to allow Pauling and his wife to go to Europe to learn the newly developing quantum theory. They left Portland by train for the east coast and Europe in March 1926. One of the curious things about their departure is that they left their not-quite one-year-old son, Linus junior, behind in the care of Pauling's maternal grandmother. They would not see him for a year and a half. Ava Helen apparently felt that having the boy in Europe might be a strain, and Pauling felt that caring for his son might take time away from his work. If anyone should have been more sensitive to the feelings of abandonment of a child by its parents, it should have been Pauling.

> In the 1860s, about fifty years before I went away to college, chemists in Germany, England, and France had decided that the atoms in substances generally can be described as forming bonds with one another. It was accepted that the hydrogen atom can form one bond, the oxygen atom can form two bonds, the carbon atom can form four bonds. For fifty years after 1865 chemists had made great progress in understanding the properties of substances by discussing various ways in which atoms can be attached to one another by these chemical bonds.

While they had discovered and classified many chemical compounds, until the first decades of this century chemists had made essentially no progress understanding what held a molecule's components together—the nature of the chemical bond. That is because until Ernest Rutherford and his students discovered the atomic nucleus in 1909, it was not clear that atomic electrons were located on the outside of the atom and thus could participate in the chemical bonding process. In 1913, after a stay with Rutherford in Manchester, Niels Bohr returned to Copenhagen where he created the Bohr model of the atom. In this model, the electrons are assigned special orbits—"Bohr orbits"—around the positively charged, massive nucleus. Given this picture, one could now imagine building the periodic table of elements by filling up the possible orbits with electrons. (It was not until the early 1920s, when Wolfgang Pauli articulated what became known as the "exclusion principle" for electrons, that a rational basis existed for limiting the number of electrons in each Bohr orbit.) The electrons in the orbits farthest from the nucleus—the so-called valence electrons—were responsible for the chemical bonding. An important step forward was taken by the very distinguished German theoretical physicist Arnold Sommerfeld, director of the Institute for Theoretical Physics in Munich. Bohr had "quantized" circular electron orbits, which Sommerfeld generalized to orbits of more complicated shapes that could interpenetrate each other. The diagram on the facing page shows these interpenetrating orbits for methane, CH_4—one carbon atom bonded to four hydrogen atoms. Notice how the six atomic electrons are shared between the carbon and the hydrogen atoms, whose positively charged nuclei have a total of six, balancing negative electric charges. Such a shared electron bond is known as a covalent bond since the valence electrons are being shared.

Pauling had decided to use his Guggenheim Fellowship to study with Sommerfeld—a wonderful choice. Not only was Sommerfeld one of the greatest teachers of theoretical physics in the world, but he was perfectly positioned to teach the "new" quantum mechanics, which was being created by people like Werner Heisenberg and Wolfgang Pauli—both former students of Sommerfeld—at the time Pauling arrived in Munich. Pauling was made quickly aware that what he had been doing in terms of the old quantum theory was—as Pauli informed him—"not interesting." Pauling soaked up the new theory, which replaced classical orbits with waves of probability. For the next decade he became a master at applying it to chemistry. This culminated in the

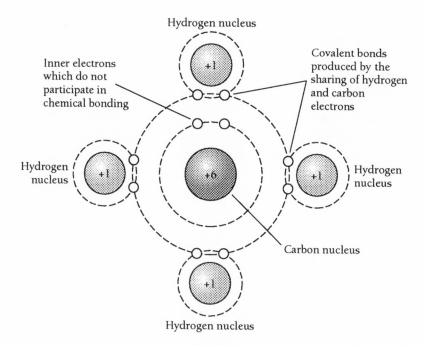

Hydrogen nucleus

Inner electrons which do not participate in chemical bonding

Covalent bonds produced by the sharing of hydrogen and carbon electrons

Hydrogen nucleus

Hydrogen nucleus

Carbon nucleus

Hydrogen nucleus

FIGURE 1.

text he wrote in 1939 entitled *The Nature of the Chemical Bond and the Structure of Molecules and Crystals*. It turned out to be one of the most influential scientific monographs of this century. When the Nobel Prize committee cited Pauling in 1954 for "studies of the nature of the chemical bond" they surely must have had this book in mind.

My wife once said to me "If that was such an important problem, why didn't you work harder at it?"

Any student of Pauling's life must be puzzled by his failure to discover the structure of DNA—the double-helix of James Watson and Francis Crick. I do not think Ava Helen's notion, that Pauling's lack of industry was to blame, gets to the heart of the matter. Before discussing this in more detail, let me explain how Pauling got into these biological matters in the first place. When Pauling returned to Cal Tech in

1927 he was made an assistant professor of "theoretical chemistry," which meant applied quantum mechanics. He rapidly moved up the ladder. In 1933, he was elected to the National Academy of Sciences—an extraordinary honor for someone so young. Meanwhile, Cal Tech had begun building up its biology department, most notably by hiring T.H. Morgan and his group from Columbia University. Morgan was one of the premier geneticists in the world, and having him at Cal Tech created an instant biology department. But the presence of people like Morgan was not what turned Pauling's focus towards biology. It was money.

During the 1930s, there was little government support for science (a condition we may now be revisiting). The major support came from private sources, the most important of which was the Rockefeller Foundation. The man in charge of distributing the Rockefeller largesse in the sciences was a physicist and statistician named Warren Weaver. By his own admission, Weaver was a second-rate scientist, but he had an uncanny ability to evaluate other people's research and create new directions, which he could do by awarding Rockefeller grants. In the early 1930s, Weaver got the idea that the techniques of physics and modern chemistry could be used to investigate biological processes. He invented the name "molecular biology" for this new activity. Between 1932 and 1959, Weaver managed to direct to this enterprise a large portion of the $90 million the Foundation spent on the sciences. One of his first grants was to Pauling—$20,000 for two years—to continue his nonbiological research.

Eventually the grant ran out and Pauling wanted a renewal. Weaver made it clear that unless Pauling was prepared to switch his interest to biological molecules there wouldn't be one. Pauling decided to study hemoglobin—to try to determine its structure and the structure of proteins in general. It was a problem he worked on for much of the next fifteen years. His first great breakthrough came in 1948. The circumstances of this epiphany are part of the Pauling legend. Pauling was in England visiting Oxford University, lying in bed recuperating from a cold, when it suddenly struck him that complex molecules like proteins must have a helical structure. This idea was based on a piece of mathematics he had learned in a course 25 years earlier at Cal Tech with the mathematician Harry Bateman. (One of Pauling's great strengths was a prodigious memory.) The figure below, which looks like a spiral of rectangles, illustrates the idea.

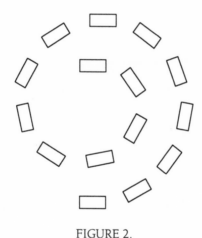

FIGURE 2.

If you take a rectangle and slide it along an axis while continuing to rotate it, you can see that will trace out a spiral or, in three dimensions, a helix. This is a special case of what is known as Bateman's theorem. In Pauling's words, it "states that the most general operation that converts an asymmetric object [our rectangle being a simple example] into an equivalent asymmetric object . . . is a rotation-translation—that is, a rotation around an axis combined with a translation along the axis—and that repetition of this operation produces a helix."

Thus, if a large biological molecule is constructed by joining to-gether a number of identical subunits (such as our rectangles) in such a way as to conserve the integrity of the subunits, then this molecule—if it has any periodic structure at all—is likely to have the structure of a helix. It is impossible to overemphasize the importance of this idea in studying the structure of these molecules. One can't "see" them in the usual sense. The X-ray data project a complicated three-dimen-sional molecular structure onto two dimensions, and without some guiding principle it is very difficult to interpret these pictures. But once one knows that they are projections of a helix onto two dimen-sions one can use these images to determine the geometry of this he-lix; that is, to solve the structure problem.

Pauling immediately decided to apply this idea to analyzing kera-tin—the protein found in nails, hoofs and horns. Some crude X-ray data existed, but when Pauling tried to fit it with a helix it didn't quite work. So he didn't publish anything for two years—until he became worried that a British group was about to publish the same idea and he would lose priority. At about the same time, new X-ray data showed that Pauling's fit had been right all along. This was the first great break-through in the determination of protein structure.

But DNA is not a protein, and, somehow Pauling decided that it needed a triple helix structure with three intertwining strands. He had used this model to analyze the structure of collagen. Unfortunately, DNA is a double helix with two strands. But Pauling became so enam-ored of his triple helix model that he couldn't let go. He was trapped in his own paradigm. Indeed, in 1953, the same year Watson and Crick published their note which created modern molecular genetics, Pauling produced an entirely wrong paper in which he claimed that DNA *was* a triple-stranded alpha helix. I think it is fair to say that this was the last deep science Pauling ever attempted. The episode took the heart out of him. Watson and Crick won the race, not only because they were very smart, but also because they were young. They had no paradigm to defend. As for Ava Helen, she had it wrong. It wasn't Pauling's lack of industry, it was his lack of youth. In science, the business of revolution is usually the province of the young.

> Before World War II, my wife hired a Japanese gardener. When the war started, all Japanese were transported to detention camps. As a consequence we lost our gardener. But before long, someone telephoned my wife to inform her of a young Japanese-American Nisei who, though already inducted into the American army, had two weeks' leave to settle family affairs and would like to take care of our garden during the interim. My wife and I belonged to a group that was protesting the treatment of Japanese people in California. Perhaps a member of that group suggested that my wife hire the young man.

> She did hire him. But he worked for only one day, because on the night of his having been hired, a rising sun and the words "Americans die, but Pauling loves Japs" appeared painted on our garage and mailbox.

> We were threatened, and the threats grew worse after word of the incident appeared in the newspapers. I had to go to Washington, D.C., on some war work; and while I was away, the local sheriff was compelled to put a guard around our house to protect my wife.

Until this episode, which affected him deeply, Pauling does not seem to have had much interest in political and social questions. His wife, on the contrary, came from a very active, politically liberal family. Her mother had been a suffragist. During the war Pauling was heavily engaged in war-related work of various kinds. It was only after the war—inspired, he reports, by his wife's statement that "The most important of all problems is that of keeping the world from being destroyed in a nuclear war"—that he began to put his full energies into political activities. Like many scientists of the time, Pauling was very concerned that the control of nuclear energy be kept in civilian hands. He spoke widely and joined a number of left-wing political groups, some of which almost certainly had Communist members. Pauling was never a Communist, but his activities soon attracted the scrutiny of the FBI. This, in turn, came to the attention of some very conservative trustees of Cal Tech, who began to pressure its president, Lee DuBridge, to keep Pauling in check—if necessary, to fire him. Pauling's activities were beginning to cost him—and Cal Tech—government funding. But DuBridge was willing to defend him so long as he would testify under oath that he was not a Communist (he did) and so long as he continued to do his job as chairman of the Division of Chemistry and Chemical Engineering and director of the Gates and Crellin Laboratories of Chemistry.

From 1952 until he received his first Nobel Prize in 1954, Pauling's travel was restricted; he was denied a passport. (Later, he would state that a significant reason he had not found the correct structure of DNA was his inability to travel to a conference in London in 1952, where some new X-ray data were shown. As I mentioned before, I do not think this was the basic reason.) Once Pauling got his Nobel, he devoted himself almost full time to political causes. He was so involved that in 1958 he was pressured to resign his administrative positions in the Division of Chemistry and Chemical Engineering at Cal Tech. In evaluating these activities one has to distinguish Pauling's generally admirable beliefs from his often self-serving and ineffective methods. For example, Pauling felt, I think correctly, that nuclear weapons testing in the atmosphere was an abomination. But he also became a public nuisance. For instance, in April 1962 Pauling, along with 48 other Nobel Prize winners, was invited to a dinner at the White House. Before going in, he insisted on joining picketers out front to protest atmospheric testing. It does not seem to have occurred to him that a more consistent form of protest would have been to refuse the invitation. Apparently, he could not resist having himself photographed in both places.

Pauling also began a series of widely publicized libel suits including one against the *National Review*, which had called him a "megaphone for Soviet policy." He lost on the grounds that he was a public figure.

By 1961, Pauling was threatening to bring suit against the *Bulletin of the Atomic Scientists*—a magazine that he had helped found. At issue was an article by the physicist Bentley Glass, who argued that both Pauling and Edward Teller—although on opposite sides of the weapons testing question—were coloring the debate by misrepresenting the data for political reasons. Teller, who tried every trick he could think of to prolong nuclear testing in the atmosphere—a position which now looks totally absurd—was wrong in failing to admit the dangers. Pauling, Glass wrote, was essentially right about the dangers, but his account of them was exaggerated and therefore vulnerable. In retrospect, Glass's analysis seems correct and Pauling's reaction to it equally absurd.

In 1963, Pauling was awarded the Nobel Peace Prize. It aroused mixed feelings, even at Cal Tech. This is clearly evident in the statement issued by Lee DuBridge: "The Nobel Peace Prize is a spectacular recognition of Dr. Pauling's long and strenuous efforts to bring before the people of the world the dangers of nuclear war and the importance of a test-ban agreement. Though many people have disapproved of some of his methods and activities, he has nevertheless made a substantial impact on world opinion, as the award clearly proves." Pauling was deeply hurt by the last part of this statement and it finally convinced him to leave the Institute—something he had been contemplating for several years.

For the next decade, Pauling was an intellectual nomad. He served as a visiting professor at the University of California at San Diego (among other positions) until 1973, when he created the Linus Pauling Institute of Science and Medicine in Menlo Park near Stanford University. He used it as a platform to promote his ideas on what he called "orthomolecular" medicine. One application was an attempt to treat cancer with massive doses of vitamin C, something that failed conspicuously when his wife died of the disease in 1981. Pauling spent the rest of his life dividing his time between the Institute and the lovely coastal ranch outside Big Sur that he and Ava Helen had bought in 1956 with the proceeds from his first Nobel Prize. This is where he died, on August 19, 1994.

Fused

For a scientist, there is nothing quite like the feeling of having one's science erupt because of a new and unexpected discovery. J. Robert Oppenheimer's celebrated—if slightly baroque—description of the emotions surrounding the discovery of quantum theory in the late 1920s captured the flavor. As he said in his 1953 BBC Reith Lecture "It was a time of earnest correspondence and hurried conferences, of debate, criticism, and brilliant mathematical improvisation. For those who participated [Oppenheimer got into the field a year or two after the principal discoveries had been made, but he knew all the participants well.], it was a time of creation; there was terror as well as exaltation in their new insight." Indeed there was. The entire edifice of classical physics was crumbling.

I had the good fortune to get into physics in the mid-1950s, when a much scaled-down version of such a breakthrough took place. In late 1956 and early 1957, following the suggestion of the Chinese-American physicists T.D. Lee and C.N. Yang, experiments revealed that parity symmetry—the symmetry between right and left handedness, sometimes called mirror symmetry—broke down in radioactive decays. It was a completely unexpected discovery, which is still not understood on the deepest level. It opened up entirely new vistas in physics and showed that nature was much less symmetrical than we had been led to believe.

I was at the Institute for Advanced Study, in Princeton, while much of this was going on. Both Lee and Yang were there as was Oppenheimer, our director. There were certainly "hurried conferences" and a stream of "earnest correspondence." I don't much remember any "terror," although Wolfgang Pauli did lose a bet. He bet that parity would be conserved. Gambling aside, there was no end of things to do. Everyone who wanted to could pitch in. Most of this work was fruitful in that, if one was a theorist, one could produce formulae that could be checked by experimenters and, if one was an experimenter, ideas for good experiments were everywhere. Morale was very high. It was a good time to be a physicist. I bring this up because on March 22, 1989, a press conference was held in Salt Lake City, Utah, which created the same kind of shock wave in the physics community. Unlike the discovery of Lee and Yang, which involved the most abstruse pure science, far removed from any practical considerations, the Utah discovery had the potential for solving our energy crisis—if not forever, at least into the far future. It held the promise of producing electricity, as the saying goes, too cheap to meter.

The two scientists who presided over this press conference, Martin Fleischmann and Stanley Pons, were electrochemists. Both specialized in electrolysis, the study of the passage of electric currents through liquids—the process that makes electroplating possible, for example. I think it is fair to say they were totally unknown to the general scientific community, to say nothing of the general public. What they were claiming to have discovered became known as "cold fusion." Fusion energy is what powers the stars. When it is uncontrolled, it is what powers hydrogen bombs. In these applications it requires a medium— a plasma—heated to temperatures of tens of millions of degrees to make it work, a process known as "hot fusion." The University of Utah scientists were claiming that, using a tabletop apparatus consisting of small palladium rods dipped into a container filled with "heavy water" arranged like a battery with the whole thing immersed in a bath of ordinary water, they had produced fusion. Why was this so spectacular? To explain, I have to say a few words about fusion.

In 1905, the same year that he published his paper on the special theory of relativity, Einstein also published the three-page addendum in which he explained the significance of the formula $E = mc^2$, which he had derived from the theory of relativity. Putting Einstein's result in modern language: an atomic nucleus, as we have discussed, is made up of neutrons (massive neutral particles) and protons (positively charged particles of comparable mass). (At the time Einstein published his paper the atomic nucleus had not yet been discovered.) These nuclear particles—"nucleons"—are bound together by the strong nuclear force. It takes a certain amount of energy to break them apart. This energy is called the binding energy of the nucleus. If one measures the mass of the constituent nuclear parts, one finds that, in aggregate, these parts are *more* massive than the nucleus from which they came. The whole, in this instance, weighs *less* than the sum of its parts. This mass difference multiplied by the square of the speed of light—c^2—is just the binding energy of the nucleus.

To take a specific example—one that plays an important role in what follows—suppose we fuse a single neutron and proton. The resultant nucleus, now composed of a bound neutron and proton "deuteron" has a mass energy of some 2.2 million electron volts less than the neutron and proton have individually. This may seem incredible, but it is true. Energy, or equivalently mass, is given up in the binding process. The convenient unit—the one I introduced earlier to measure these mass differences—is the electron volt. Even the energy of a million electron volts is tiny compared with the energy needed to illuminate a light bulb. But when billions upon billions of these nuclei undergo a fusion process, it adds up. Witness the hydrogen bomb. When the neutron and proton fuse to make a deuteron, the excess binding energy is released in the form of an energetic radiation quantum—a gamma ray—which is observable. One can measure its energy and test Einstein's formula. Even more energy is released if the deuterons themselves fuse to make a still heavier isotope of hydrogen or a light isotope of helium. In ordinary seawater, about one in 6,700 hydrogen atoms has a heavy-water nucleus. It appears, on the face of it, that one could extract almost limitless fusion energy from the sea.

But there is a catch. (There is always a catch.) Particles with like electric charges repel each other. For example, two heavy-water nuclei—deuterons—repel each other since each has a positive charge. To overcome this, the charged nuclei, if they are to fuse, must be flung together at high speeds. This is why enormous temperatures are required to attain that result in the hydrogen bomb or the Sun, to name two common examples. Fusion in a hydrogen bomb is triggered by first causing the explosion of a Hiroshima-like fission bomb. But since the war, billions of dollars have been spent on attempts to reproduce these extreme temperature conditions in the laboratory. To give some idea of the scale, the largest test hot fusion machine—the ITER (International Thermonuclear Experimental Reactor)—is now under construction. It involves scientists from all over the world and is expected to take some thirteen years to build at a cost of $7.5 billion. The annual budget for hot-fusion research in the United States is about $350 million. I have often visited the Princeton Plasma Fusion Laboratory where a smaller version of the ITER—the TFTR—has been constructed. It employs hundreds of people in a wide variety of disciplines. In their press conference, Pons and Fleischmann were claiming to have accomplished the end result of all this in an experiment that could almost have been done in a high school chemistry laboratory for perhaps tens of thousands of dollars. It was a stunning feat, if true. But was it true? First, some background:

Martin Fleischmann was born on March 29, 1927, in Karlsbad, Czechoslovakia. His family emigrated to England just before the war. He took his degree in chemistry at Imperial College in 1951 and rapidly acquired a reputation as one of the most innovative electrochemists in the world. To attain such a reputation, he seems, serendipitously, to have followed the advice that a friendly senior colleague gave me when I was just starting my career in physics. He said the reputation I should strive for was to be brilliant, but erratic. That way, he pointed out, I would get good jobs but would never be asked to serve on committees. Fleischmann certainly had the reputation of being brilliant, and many of his ideas were erratic. But enough of them weren't, so he became a professor at the University of Southampton and a Fellow of the Royal Society. He took early retirement from the university in 1983 at the age of 56. He met Stanley Pons in the mid-1970s when Pons came to Southampton to do his doctorate. He had been born in 1943 in Valdese, North Carolina, where his father owned several textile mills. After graduating from the University of Michigan, he worked for a time in the family business, but in 1975 he decided to finish his degree work in Britain. He and Fleischmann hit it off instantly, finding they had many common interests, including cooking and skiing. When Pons moved to the University of Utah, in 1983, he asked the now retired Fleischmann to visit on a regular basis.

Fleischmann had long been interested in a peculiarity of the metal palladium. It soaks up hydrogen—both heavy and light—like a sponge. The idea occurred to him that if one immersed palladium in heavy water it would take up massive amounts of deuterium. The deuterium in the palladium might, he reasoned, pack together so tightly that fusion reactions would take place. The date of this epiphany is a little unclear, but it would appear to have been sometime in 1987. He made the decision to test this idea as he and Pons were hiking in Mill Creek Canyon in Utah. In fairly short order, the two men constructed the palladium cells and began measuring heat and radiation emanating from them. They noted an apparently significant increase in both, after the palladium had been loaded with deuterium. Soon after making this observation, Fleischmann had to undergo major surgery and could not take up the project again until he returned from England a year later. However, prior to his return, while Pons's son was minding the palladium cell, an explosion evaporated a 1-centimeter-cubed block of palladium. Since they were anticipating nuclear fusion, they decided that this must have been it—a miniature nuclear explosion. Upon Fleischmann's return they began a series of experiments to see if this interpretation could be sustained.

Meanwhile, an independent but related enterprise had been unfolding at nearby Brigham Young University in Provo. For some time, a physicist there named Steven Jones had been interested in a form of cold fusion that we know works. The electron, the particle that circulates the atomic nucleus and is responsible for the atom's chemistry, has, for reasons we don't understand, a more massive analog. For historical reasons, it has been given the unfortunate name of the mu-meson or muon. This heavy *doppelgänger*, although some 200 times more massive than the electron, has the same electrical properties as the electron. It can be induced into normal atoms, thereby replacing the electrons. When this is done, say, for hydrogen, the resulting atom is about 200 times smaller than normal hydrogen. The muon is some 200 times closer to the proton nucleus than the electron. This dwarf hydrogen—or muonic hydrogen—can get closer to its neighbors than standard hydrogen. When this happens, fusion can take place. The muon then escapes to initiate another fusion process.

One might think that muon-catalyzed fusion, as it is known, would be the solution to all our problems. The trouble, however, is that the muon is unstable and decays in about a microsecond back into an ordinary electron and two elusive neutrinos. This happens too quickly for the muon to catalyze enough fusions to produce any significant amount of energy. Nonetheless it is a fascinating process and has been observed repeatedly in the laboratory.

Jones got the idea that these fusion rates could be increased if the hydrogen was put under pressure. He even conjectured that an important mechanism for heating the interior of the Earth might be muon-catalyzed fusion taking place in compressed metals like palladium. With several collaborators, he began a series of experiments that bore a family resemblance to those of Fleischmann and Pons. The object: to study possible fusion reactions in palladium cells in which deuterium was under high pressure. Jones did not believe (and never claimed) that such a mechanism had any application towards solving our energy problems. He was simply doing research in basic science. In the resulting melée, Jones's name got entangled with those of Pons and Fleischmann, which was unfortunate and clouded the issue. By the beginning of 1989, Jones was planning to go public with his results, which Pons and Fleischmann became aware of. But their perspective was entirely different. They thought they had solved the energy problem by devising a method that could probably be patented, making them—and incidentally the University of Utah—rich beyond avarice. Hence, the hastily called news conference of March 23rd. They wanted to stake their claim.

Most physicists who were active in 1989 probably remember where they were when they first heard the astounding news that our energy problems had been solved by two chemists from Utah. I was in New York with a group of physicists from Columbia University, including the aforementioned T.D. Lee, having a Chinese lunch in a restaurant on upper Broadway—a local tradition. Someone broke the news and we talked of nothing else. There was much frantic scribbling on paper napkins until it became clear that something about the Pons–Fleischmann data did not add up. Where were the neutrons? When two deuterium nuclei fuse, one inevitable outcome is the production of a light isotope of helium and a neutron. The production of neutrons is a sure sign that the reaction is nuclear fusion. Pons and Fleischmann were claiming that when they observed the additional heat from their palladium cells, they also observed something like a 50 percent rise in the background radiation. But to be consistent with the amount of heat they were seeing, they would have to have observed a *billion*-fold rise in the neutron background. In fact they would have been dead—irradiated. Thus, believing in cold fusion required believing in two miracles—that deuterium under pressure fused at the rate claimed, and that somehow there were no neutrons.

For many people, myself included, this was too much. I had no idea what they had done wrong—I still don't—but I decided that whatever it was, it was *their* problem and I promptly lost interest. So much was at stake, however, that a kind of feeding frenzy developed in the general community. Among other things, there were national security implications. Here is the logic: Another possible outcome of deuterium fusion is the production of a still heavier hydrogen isotope called tritium—one proton and two neutrons. This is used as an essential fuel in the hydrogen bomb and is not easy to manufacture. It does not occur naturally to any great extent because it has a half-life of about 12 years before it decays into helium. But if the Utah chemists were right, tritium could now be manufactured by high-school students, either conveniently at home or in their school chemistry laboratories. The prospect was so alarming that Admiral Watkins, under a directive from President Bush, formed an emergency panel to study the situation. This was a month after the initial press conference, by which time the whole scenario was beginning to unravel.

It is very difficult to do a scientific experiment to search for a null effect. There is always the possibility that a small effect has been missed or masked. Until people began doing experiments to try to reproduce the results of Pons and Fleischmann, they had not fully realized just how difficult it was to detect a small flux of neutrons. The conventional detectors are designed to deal with massive radiation. To this day, as far as I know, it is not clear that even Jones, who was poised to see a small flux of neutrons from the fusions that were presumably taking place in his experiments, really saw any. What became clear was that, try as they did to replicate Pons and Fleischmann's experimental setup, none of the other laboratories qualified to detect neutrons saw any. But this does not seem to have deterred Pons and Fleischmann. Indeed, their high-water mark occurred at a meeting of the American Chemical Society on April 12, 1989—three weeks after their press conference. The atmosphere was, to put it mildly, intense.

According to an eyewitness description, Pons was mobbed like a "rock star." The president of the Chemical Society at the time, Clayton Callis, the same observer noted, opened the special session on cold fusion and remarked that the physicists' efforts at hot fusion using tokamaks* and lasers "were apparently too expensive and too ambitious to lead to practical power." He added "Now it appears that chemists have come to the rescue." At this, the crowd reportedly burst into laughter and applause.

Within a month, the chemists had little to laugh about. Cold fusion, as far as most of the scientific community was concerned, was essentially dead. Indeed, the very simplicity of the Pons–Fleischmann setup, the fact that it was a tabletop experiment, led to its rapid denouement. To make matters worse, Pons and Fleischmann were not very helpful. At first they offered to move some of their equipment to the laboratory at Los Alamos, which had the wherewithal to measure the neutrons. In the end, they never did, citing possible patent conflicts. In fact, in April 1990, Gary Triggs, Pons's lawyer, threatened to sue the authors of an article that appeared in the British journal *Nature* refuting the pair's claims. In the end nothing happened, but it was an unprecedented attempt to interfere with a scientist's right to publish. When it was pointed out to Pons and Fleischmann that the gamma rays in one of their experiments had the wrong energy to have come from fusion, they obligingly moved its energy. It is difficult for me to imagine, since I don't know them, what they thought they were doing.

It is fashionable now to point out what appear to be ethical lapses among scientists, as if such lapses were the norm. What this episode shows is just how exceptional such behavior is and how rapidly it can be exposed once the community focuses on it. It took some two

*This toruslike contraption, of which Andrei Sakharov was one of the original designers, appears to be the most promising tool for actually making hot controlled fusion work.

months from the original press conference before most of the scientific community agreed that cold fusion was a fantasy. I do not know if hot fusion will ever solve our energy problems. But I recall an exchange I once had with the great historian of religions Harry Wolfson. When he retired from Harvard in the 1950s he was allowed to keep his office in the bowels of Widener Library. He had been one of my father's teachers so I used to visit him fairly often. He had a Jewish accent one could cut with a knife. At the time he was working on the history of what he called the "Choich Fadders." On one occasion I brought along a used book I had just bought cheap, only to notice that some of its pages were missing. He looked at it and said, "If you pay for a boggin, you get a boggin [i.e., junk]." In the energy business there are no bargains.

———

Several years have now passed since the press conference of Pons and Fleischmann. There still seem to be True Believers. As I note in another essay in this collection, my teacher Julian Schwinger was one. My very old friend Arthur Clarke, the science fiction writer is another. Every few months I receive a report from him. Here is a recent one, written in November 1995: "I'm still keeping my fingers crossed on the 'new energy' front. You may have heard that Pons and Fleischmann have just been issued a European patent. However, I think they have been bypassed by history—there have been some stunning developments elsewhere, and even the U.S. patent office has had to capitulate in the face of incontrovertible demonstrations." It reminds me of a story my father told about his experience as General Eisenhower's advisor on Jewish displaced persons (DPs) after the war. These Jews, who had emerged from the concentration camps, were living in Spartan camps in Germany run by the U. S. Army. An individual DP could get somewhat better treatment if he or she had a vital trade or skill. A call went out for cobblers.

Suddenly a plethora of "cobblers" appeared whose skills had nothing to do with shoes. The army got wise, and these "cobblers" were demoted to their previous status. But then my father received a delegation from the displaced persons whose spokesman asked, "Now that you have taken away the cobblers, who will fix our shoes?"

Smuggler

By the summer of 1991, it had become clear even to longtime visitors like myself—just by reading the local newspapers—that the community of Aspen, Colorado, and the Environmental Protection Agency in Washington were on a collision course. This seemed extremely ironic since no community that I have ever lived in is more preoccupied with its environment than Aspen. You cannot remove the slightest cottonwood tree from your front yard without its becoming a subject for protracted debate at a city council meeting. Moreover, the unlikely catalyst for this confrontation with the EPA was a 38-year-old registered nurse at the Aspen Valley Hospital named Patti Clapper, who, as both a nurse and a mother, had as much concern for environmental protection as anyone in the valley. Nevertheless, anyone who reached Patti Clapper's answering machine that summer heard a poem written by her husband Tom, a ski instructor in the winter and stonemason in the summer. The poem, which was recited on the machine by Patti, went:

> Roses are red.
> Violets are blue.
> We're fighting the EPA,
> And we hope you are too.

How Patti Clapper came to lead an ultimately successful citywide fight against the EPA is one of the most remarkable local sagas of recent years, and one that has changed the way that agency deals with the Western mining states.

It all began rather innocuously in the spring of 1981. A graduate student at Colorado State University named David Boon decided to test the fertility of the soil in and around the Smuggler Mobile Home Park, a small, compact neighborhood of mobile homes at the foot of Smuggler Mountain, on the northeast side of Aspen. There are some 80 mobile homes there, which in fact don't look very mobile. A few have two-car garages. Tom and Patti live with their two young children in one of these mobile homes. (Patti prefers to call them "modular homes.") They bought it in 1987 for $40,000, a price that included its tiny plot of land. The Clappers' home, which is now about 25 years old, is mobile only in the sense that they have the various devices needed to move it once the concrete foundation has been torn away. It has three bedrooms (one just large enough to contain a crib), a kitchen, a bath and a living room. And there is a garden in which the previous tenants grew fruit and vegetables—a common practice in the mobile home park. One reason Boon was studying the soil was to help the Smuggler people with their agricultural activities. He also had an academic interest in lead and cadmium in the soils near old silver mines.

There are two old mines just a few yards from the Smuggler Mobile Home Park. One is the Smuggler Mine, which began producing silver in 1880. It is currently operated by a man named Stefan Albouy, who does some occasional mining and takes people on tours of the mine. The other, which started up about the same time, was called the Mollie Gibson; many things around Aspen are named after Mollie Gibson, who according to legend was a noted lady of the evening in the mining town of Leadville. In reality, she seems to have been the wife of a Colorado judge as well as the sister of the mine's original owner. As photographs from the heyday of silver mining show, the area around what is now the Smuggler Mobile Home Park looked like the industrial corridor of New Jersey—all smokestacks. Both mines stopped large-scale production toward the end of the First World War, by which time they had yielded millions of dollars worth of silver ore. But their activities left a lot of residue in the soil, and that is what Boon was looking for.

To set the scale, the EPA considers 500 parts per million an acceptable level of lead in most soil samples. For reasons of practicality, the permissible level around Aspen is 1,000 parts per million. If the standard were more rigorous, Aspen, and indeed much of Colorado, would be unable to meet it. However, Boon discovered places in and near the mobile home park where the lead levels were over 20,000 parts per million. The first suggestion that something ought to be done about this situation was a press release issued by the Colorado State University extension service on August 28, 1981. It read "Caution is advised to any individual or family whose garden is located near a mine site, for the soil may have toxic qualities. Remember, children should not be fed these vegetables, due to the risk of lead poisoning."

Boon's findings came to the attention of Tom Dunlop, the environmental-health officer for Pitkin County and Aspen, which is the county seat. As Dunlop later explained to me, "I didn't know David Boon from Adam." So he repeated the tests and got the same results. He immediately issued public notices requesting that Smuggler residents stop eating vegetables from their gardens and keep their children from playing in old mine tailings. He also informed both the city council and the county commissioners of the situation. At that time the mobile home park was part of the county but not the city—a circumstance that changed a few years later, with dramatic consequences for the events of the summer of 1991.

It is Dunlop's impression that when he first alerted the city and county authorities, there were five or six hundred people living near the old Smuggler Mine—some in the mobile homes and some in a modest middle-class development called the Hunter Creek condominiums. But the town was expanding rapidly in the area of the mine. It is one of the few neighborhoods where people who work in Aspen can actually afford to live, so the population by now has more than doubled. A newer development, known as the Centennial condominiums and consisting of some 240 residences (now completed) was then under study for approval by the county. The man behind Centennial was Sam Brown, who had the reputation, as a former county commissioner put it to me, of being "a good developer"—one who was deeply committed both to environmental issues and to middle- and low-income housing. (Brown was an organizer of the first Earth Day and was in charge of Vista and the Peace Corps during the Carter administration.) The county commissioners felt they had a lead problem in the area, and they were worried about letting more families move up there unless some remedial work was done, such as covering the most heavily contaminated places with fresh soil.

As it happens, a year earlier Congress had enacted the Comprehensive Environmental Response, Compensation, and Liability Act, which created the EPA's Superfund program. The Superfund, established with an initial outlay of $1.6 billion, was to be used to clean up toxic-waste sites around the country. The program came to the attention of a young county engineer named Pat Dobie. Dobie felt that if Aspen could qualify as a Superfund site this might help finance the remediation, which the county had estimated would cost about $500,000. Back then, none of the county commissioners knew a great deal about how the Superfund program worked, but Dobie had heard about a provision concerning Potentially Responsible Parties, or PRPs. These were entities from which the government could recover the cost of the cleanup—or at least part of the cost.

It was not clear to the Aspen officials then what fraction of the total the PRPs, if they could be identified, would pay and what fraction the government would pay. It seemed to Dobie and the commissioners that some of the PRPs in Aspen's case were quite wealthy—for instance, the descendants of the principals of the Smuggler Durant Mining Corporation, the original operator of the Smuggler Mine. (Indeed, the list of PRPs eventually grew to include the Atlantic Richfield Company, which had leased mineral rights to some of the mine tailings in the late 1940s.) In fact, the commissioners reasoned, the PRPs might well pay for their share of the cleanup and for a whole lot of new landscaping besides. Hence the commissioners wrote a letter to the EPA, inviting the agency to turn its attention to Aspen. Not long afterwards, the EPA sent a team to begin studies of groundwater, surface water, and air quality, which might lead to the area's qualifying as a Superfund site. The decision would be made on the basis of a point score: the minimum number of points required for a site to be put on the Superfund's national priorities list was 28.5. Aspen scored what Dunlop referred to as "forty-four and change."

The tests that the EPA did at that time have been disputed in Aspen ever since. No lead was then—or, indeed, ever—found in the water supply. A cadmium level of 13 parts per billion was found in a private well; 10 parts per billion is the EPA's allowed limit. A couple of months later, a second sample from the well showed 8 parts of cadmium per billion. But the scoring had already been recorded and was not, according to the EPA, reversible. [The real problem with the water in the beautiful mountain streams near Aspen is not cadmium or lead, but the presence of *Giardia lamblia*, a parasite that causes serious intestinal disorders. It is carried by beavers and other animals and has rendered the water in most of these streams too dangerous to drink without first treating it with something like iodine.] The air-quality data were taken from tests that Dunlop had already performed; they showed that the lead particulates were lower than the national standard by roughly a factor of ten. Nonetheless, the EPA decided that the air was a source of concern.

By 1984, when the community began to challenge the EPA, Pitkin County had discovered that it was *itself* one of the Potentially Responsible Parties, so its taxpayers were going to have to pay some indefinite part of the cleanup. In June of that year, representatives of the EPA visited Dunlop in his office in the Aspen City Hall to inform him that an area of some 75 acres at the base of Smuggler Mountain had been designated a Superfund site and that the cost of the cleanup, which would now involve moving a layer of the contaminated soil, would be somewhere around $1.5 million. At that point the county and the other PRPs—which by now included Sam Brown's Centennial Company, Stefan Albouy and the Smuggler Racquet Club, a modest tennis club next to the mobile home park—decided to hire a firm of environmental engineers named the Fred C. Hart Associates to reevaluate the tests. Using data supplied by the county and the EPA, the firm retested the site and arrived at a score of 12.96—way below what was needed to qualify as a Superfund site.

By 1986, when the site was formally put on the national priorities list, the EPA had revised its point score to 31.31. The air-pollution number was reduced to zero (since Dunlop's readings had been taken from the county courthouse half a mile from the site, and presumably reflected conditions—such as automobile traffic—in town). The surface water was found to have negligible contamination, but the groundwater contamination was considered serious enough to justify the designation. In the spring of 1985, the EPA drilled seven additional wells to test the site for contamination, at a cost (according to the *Aspen Times*) of $18,000. One of the wells collapsed, one was abandoned and two came up dry, even after the engineers dug down some 65 feet. (To find water, the old silver miners often had to dig as deep as 160 feet.) The remaining three showed no lead contamination, and no subsequent test has ever shown that the water on the site contains any measurable amount of lead, though appreciable levels of cadmium and zinc have been found.

In September 1986, the EPA announced that it had found a site to dump the contaminated soil. This was what is euphemistically called the Mollie Gibson Park, a two-acre boulder field above the racquet club, which, even without the influx of contaminated soil, looked more like a dump than a park. The agency also announced that it might have to dig down as much as *four feet* to remove all the contaminated soil, and might have to move the mobile homes to do so.

I think it is fair to say that, at the time, few people in town were paying much attention to all of this. In fact, Dunlop told me that whenever he spoke about the impending cleanup—and what it really entailed—before the city council or the county commissioners, hardly anyone showed up. One wonders how many people might have turned out if the EPA had proposed to dig up four feet of soil in one of Aspen's many fashionable neighborhoods—for example, Starwood, a 1000-acre tract on a mesa with its own private guards, where people like John Denver and Rupert Murdoch had homes; or Red Mountain; or the West End, where the Aspen Institute for Humanistic Studies is located. I am quite sure that if the EPA had proposed to dig up one of these neighborhoods it would not have taken until 1988—when Patti Clapper began her rebellion—for there to have been major political repercussions.

The Clappers moved into the Smuggler Mobile Home Park in 1987, aware that it was an EPA Superfund site but knowing little about the consequences of such a designation. It never seemed to them even a remotely dangerous place to live. Tom Clapper, a third-generation Aspenite, had played among the mine dumps as a child but never suffered any of the effects of lead poisoning. In the spring of 1988, he went to a public meeting with officials from the EPA and was startled to learn that he—and all the other residents of the mobile home park (along with the rest of the people living on the Superfund site, which had now grown to 110 acres)—could also become Potentially Responsible Parties. This meant they could be held liable for part of the cost of the cleanup, even though some of them (as in the case of the Clappers) had moved in only a couple of years earlier.

However, the EPA had come to the realization after some site visits that moving the mobile homes and digging down four feet was probably unreasonable, and they decided instead to leave the homes where they were and to dig down only two feet. Even that would involve moving as much as 120,000 cubic yards of soil, which would amount to 12,000 truckloads! At this point the EPA decided that the Mollie Gibson Park would not be big enough to hold all the contaminated dirt and that some of the land on the adjacent racquet club would have to be used as well. The original maximum cost of around $500,000 had now risen to at least $5 million.

By August 1989, it seems to have occurred to officials at the EPA that the project was getting out of hand. So in something that became known as the Denver Agreement, the six-state regional administrator of the EPA, James Scherer, agreed to reduce the depth of the soil to be removed to one foot and then lay a carpet—a device that the agency referred to as a geotextile cover—made of a feltlike material to protect new soil from any contaminated soil remaining beneath it. The EPA also announced that it would be back in five years to reevaluate the remedy. The community was shocked when, at a meeting that October in Aspen, the EPA's representatives announced that it would cost $500,000 to do the legal work required to put the Denver Agreement into effect. By this time, the agency had spent more than $1.5 million on the project, and nothing had been cleaned up.

In the spring of 1990, Patti Clapper decided to take matters into her own hands. She had become seriously concerned, both as a registered nurse and as a mother, about the dangers of lead. "I talked to the pediatricians, and the radiologists, and the internists and the orthopedic surgeons to see if they had ever seen anything that looked like cases of lead poisoning in the area," she told me. None of them had. Patti's next step was to call Dr. Chris Weis, one of the EPA's toxicologists in Denver, to ask for information. "He made me feel really stupid, and I hate that," she recalls. "He told me, 'It's your responsibility to your community to educate yourself on the risks of lead.' I said, 'Fine.' He said, 'I have volumes of information.' I said, 'Terrific. Send me some.'"

Among the documents that Weis sent her were three studies on lead poisoning, none of which seemed relevant to the circumstances of the mobile home park. One was a 1960 study done in Boston that showed black children to have a higher incidence of lead poisoning than white children. Another was a study in the early 1970s that found poor children to be more susceptible to lead poisoning than wealthier children. The third study showed a high incidence of lead poisoning in children from backgrounds where the level of education was low.

"So, therefore, if you were poor, dumb, and black, your chances of having lead poisoning were much greater," Patti told me. "I couldn't really tell how that fitted in here. Technically, in this community of multimillion-dollar homes, those of us who live in the mobile home park are poor. But on the average for the nation we're not. We're middle class. My two nearest neighbors have master's degrees. My husband has five years of college; he has two degrees. I wouldn't say that we are uneducated. I don't really care if people are black or white. It doesn't make any difference to me. But we don't have, at the moment, any black families living in the mobile home park." (Very few blacks live in any part of Aspen.)

At that point, Patti decided that she needed to do her own research, so she began calling around. She discovered that a study of the health risks from old mine tailings had been done in Leadville, some 30 miles east of Aspen, and that similar studies had been conducted in Telluride, Colorado; Park City, Utah; and Kellogg, Idaho. The studies all had one thing in common: the finding that there was only a weak correlation between the amount of lead present in those mine tailings and the levels of lead in blood samples taken from people who lived on the tailings. People living near old mine sites that had been equipped with smelters were found to have elevated lead levels in their blood, but where there had been no smelters the lead levels were closer to normal. There were no smelters on the Smuggler and Mollie Gibson mine sites; the smokestacks that had made the area look like parts of New Jersey were from generators that pumped water out of the mines.

Patti became convinced that it would be riskier to clean up the Aspen site than simply to leave it alone, since cleaning it would mean moving tens of thousands of cubic yards of dirt with its attendant dust. She began writing letters to government officials and tried to persuade the community that the EPA had embarked on a costly folly—to little avail. She also got in touch with the Colorado Department of Health and spoke to a woman there about testing the residents on the site for lead poisoning. The woman told her to write a letter to the department asking that a public-health study be conducted—a study that, as it turned out, the department had wanted to do for some time.

Patti's letter was forwarded to the Agency for Toxic Substances and Disease Registry in Atlanta, and the agency funded the first serious human-health study on the Superfund site. It was carried out in August and September of 1990. A hundred and twenty-five volunteers on the site were tested for blood-lead levels; the average was found to be 2.77 micrograms per deciliter for children and 3.4 micrograms for adults. The Centers for Disease Control define a toxic lead level in blood as anything over 20 micrograms per deciliter of blood, and consider a level of 10 micrograms per deciliter to be a matter of concern; the national average is from four to six micrograms per deciliter. One teen-age boy tested high, but he later admitted that he spent a lot of time riding his motorbike back and forth across the mine dump kicking up clouds of dust. The EPA's target, after remedial action was taken, was to have 95 percent of the residents with a blood level of less than 10 micrograms. Not only were the lead levels of the residents on the site already about one-third of that, but they were less than the national average. However, the EPA decided to discount these tests arguing that "short term measures of exposure for a small population of individuals known to be exposed to high soil lead concentrations" were unreliable. Rather than proposing a continuing series of tests, which would have been both prudent and economical, the EPA decided to go ahead with the cleanup anyway.

I first began paying any attention to this in the spring of 1991. Up to that time I was not entirely sure exactly where the mobile home park was and I didn't know anyone who lived there. But that spring, if you read the local newspapers at all, you simply could not ignore the fact that something quite bizarre seemed to be happening in Aspen. By May 1991, the time I arrived for my annual summer visit, there had been some further developments, which I learned about once I started putting the pieces together. In April, a four-party contract called the State Superfund Contract had been signed by the city, the county, the state and the EPA authorizing the cleanup to begin on the first of August. The city had annexed the base of Smuggler Mountain in June 1988, so it now had to be a party to any dealings. This was unwelcome news to the EPA, since it had no financial leverage over the city, which was not deemed one of the Potentially Responsible Parties. The cost estimate of the Smuggler remediation, including some $3.5 million in administrative expenses, had now risen to $13.7 million. Nationally, at the time of the signing, the EPA had designated some 1,200 Superfund sites, and had cleaned up about 60. A second development was now that the Superfund site had become part of the city, the EPA would hence-

forth have to deal with the elected Aspen City Council, a redoubt-
able group made up of highly educated professional people from
all walks of life. They devote at least 10 to 20 hours a week to the
job, for which they are paid about $5,000 a year. By virtue of this
they become the targets, as one of them told me, of "two or three
million dollars apiece in lawsuits."

But perhaps the most important event of that spring regarding
the Superfund site, was an appearance that Patti Clapper and Tom
Dunlop made in late April at a weekly men's lunch club, whose members were some of the most influential people in Aspen. "It scared me
to death," Patti told me. "I am not a public speaker. I get nervous. But
I decided to do it, what the heck!" Two weeks earlier Patti had received a copy of a study done in Butte, Montana, by Dr. Robert
Bornschein, director of environmental epidemiology at the University of Cincinnati Medical School's Department of Environmental
Health. The department, it turned out, had been studying lead and
lead toxicology since the 1920s and Bornschein was regarded as one
of the world's foremost lead toxicologists. Butte is built on dozens—
if not hundreds—of copper mines, and one might well assume that
the whole town would be toxic. Indeed, the lead levels in the soil are
fairly high; however, the blood tests told an interesting story. In the
oldest parts of the city, where smelters had been active and houses
had lead plumbing and were painted with lead-based paint, the blood
levels were indeed elevated. But a trailer park in Butte was built right
next to some old mine tailings. No smelters had been in operation
there. The mobile homes were new, and the paint used on them was
lead-free; the blood-test levels of their occupants were low. To Patti,
the significance of this for Aspen was obvious. As she put it, "the lead
in these smelter-free mine tailings was not bioavailable. You could
serve it for lunch."

One of the organizers of the weekly men's get-together was a 49-year-old dentist named Terry Hale, the one who had invited Patti Clapper and Tom Dunlop to speak. Hale told me that after he heard Patti talk he went up to her and said "'You're holding back on us a little. You're giving us a one-sided deal. This thing couldn't be as ridiculous as it sounds.' She said, 'It is,' and she rather challenged me to take a look at it myself. So I did. I made a few phone calls to the Colorado School of Mines, and I spoke with some of their mine toxicologists. They confirmed what Patti told me." A few evenings later, Hale was stricken with a kidney stone. To relieve the pain, he went for a walk and ended up in the courthouse, where a meeting was going on between representatives of the EPA and the Smuggler mobile home people. Public comment was requested, and, in excruciating pain, Hale stood up and made an impassioned speech. That night he went to the hospital, but the next morning he felt well enough to go back to work. At ten o'clock, he got a phone call from the EPA saying it was flying one of its toxicologists in to talk to him and anyone else he would care to bring along. Hale is a considerable public orator, and the EPA must have realized that there was now going to be a real fight. Hale assembled a group of Aspen residents he considered to be knowledgeable in government, science and medicine to question the toxicologist; the group was "very unhappy with the answers," he said. Later, Hale, through his medical contacts, found his way to Bornschein, who then came to Aspen for two days—at his own expense—and described his Butte studies to a number of council members. He noted that the average blood levels in Cincinnati—his home base—were about 17 micrograms per deciliter, a typical figure for several eastern cities. The residents of the Smuggler Mobile Home Park had blood-lead levels about one-sixth as high.

In addition to Patti and Terry Hale, I also talked a good deal to Frank Peters, who is in the hotel business in Aspen and had been a member of the city council since June 1989. Peters had a somewhat less polemical view of what was going on at the base of Smuggler Mountain than either Patti Clapper or Terry Hale. He had been elected, at least in part, by a bloc of voters from the mobile home park—recently enfranchised by the annexation. His concern was to find out what his constituents wanted. He discovered that their main wish was to get the EPA off their backs for reasons that had become financial. Banks were now refusing to make loans collateralized by real estate on the Superfund site, since they feared that they too might become PRPs by virtue of their stake in those properties and, as such, might end up paying part of the cleanup cost. Moreover, people on the site could no longer count on selling their homes. Who wanted to buy into a Superfund site and risk becoming a PRP? Peter was willing to concede that there might be a potential health risk from the high concentration of lead in the soil, but he felt that it could be dealt with by what the EPA called "institutional controls." This concept surfaced at the time of the Denver Agreement in 1989. The idea was that, a number of rules and regulations—institutional controls— would be implemented to insure that whatever remedy effected for the cleanup would remain in place. They would, for example, limit the amount of dirt that could be moved in constructing a new house, or limit the depth to which one could dig in planting a garden.

Everyone agreed that institutional controls were an essential part of the process, but nearly everyone—including Peters—who had looked into the matter also agreed that there was no present danger from the lead. According to Bornschein, it is locked into a mineral matrix that appears to block its bioavailability. If swallowed—as in, for example, the vegetables from the gardens around the mobile homes, it passes harmlessly through the system. Moreover, the particles are too large to lodge in the lungs. Therefore, although the lead levels are very high in the soil, lead does not show up in the residents' blood. The lead in automobile exhaust, smelter slag and paint consists of minuscule unbound particles, which can be both absorbed by the lungs and digested.

What worried Peters was that without institutional controls, people on the site would move a few hundred cubic yards of toxic soil—to construct a garage, say, or put in a tennis court—and this would break up the particles and put toxic dust into the air. He had no confidence that the EPA could, as it had proposed, move 60,000 cubic yards of soil without releasing all kinds of dust. He was also worried that once contracts for the cleanup were let, the pressure of keeping within the agreed cost limits would tempt truckers to cut corners. The EPA was still planning to have most of the dirt trucked into the Mollie Gibson Park, where it would create a mountain of toxic soil 40 feet high! The agency had no plans to do blood-level tests during the two-year period while all of this was supposed to be going on. To protect itself, the community would have to conduct its own tests. During the summer there was a meeting between representatives of the EPA and the city council. The city attorney raised the question of what the agency's position would be if the city council passed its own institutional controls, without accepting the rest of the EPA's program of remediation. An EPA lawyer, said that, in that case, the EPA would sue the city.

Given everything, the opponents of the cleanup—now most of the town—decided that their most pressing task was to keep the bulldozers from attacking the base of Smuggler Mountain, which was supposed to start imminently. Various tactics were considered. One was to enlist the aid of Colorado political figures like Colorado Senator Tim Wirth. That was eventually done, and Wirth came to Aspen on two occasions to visit the site. Another tactic, which would have shut the process down cold (at least temporarily), would be to get 1,500 signatures on a petition to hold a countywide referendum on whether the county commissioners should be required to renege on their participation in the State Superfund Contract. If the referendum passed, the county would have to back down on its offer to use the Mollie Gibson Park as a dump site, or in its commitment to institutional controls for at least a year—the span of time before the county could take action again. The EPA would thus be forced into litigation before the bulldozers could roll as scheduled.

By this time, the EPA seems to have gotten the message that things were rapidly coming unglued in Aspen. On July 8th, James Scherer, its six-state administrator came to Aspen to appear at a meeting of the city council. It was none to soon. The council was threatening to table the institutional controls, and John Bennett, Aspen's mayor, had gone as far as to say that if the EPA's bulldozers showed up at Smuggler, he would stand in front of them. I attended the meeting, which was open to the public. Scherer turned out to be a kindly looking grey-haired man who seemed genuinely sorry for all the trouble the EPA's actions appeared to be causing. However, his general message to the council was that the city should simply get this unpleasant business over with as soon as possible, as if it were a visit to the dentist. At one memorable point he said the Superfund law was "East-dominated" and "not designed for the West," and he added "I don't like the Superfund law. I hate it." But he also pointed out that his job was to enforce the law. He told the council that to get Aspen delisted as a site against the will of the EPA would require a specific act of Congress.

Events moved rapidly after the July council meeting. Unmoved by Scherer, the council members decided to table the matter of institutional controls for six months. That in itself would have been enough to freeze the process, but at the urging of the council, the county commissioners voted to rescind their own agreement on institutional controls. The referendum had gathered almost all the 1,500 necessary signatures, but the council and the county commissioners took these actions because a favorable outcome in the voting—a nearly sure thing—would have tied the hands of the county for a full year. If the EPA were to come back with a reasonable proposal before the year was out, the county could not have acted on it. Impressed by what he had seen (and, no doubt, by the political clout of the Aspen community), Senator Wirth wrote an open letter to the EPA, urging moderation. He also had a private talk with EPA officials in Denver.

This seemed to have had an effect. On August 7th, James Scherer wrote a letter to Senator Wirth announcing that the EPA would postpone any remediation of the site for at least two years. The morning after this announcement, I ran into Terry Hale in a dry-cleaning establishment. He was pleased but not jubilant. In the first place, the EPA had not delisted the site, so the people living there were going to suffer all the financial inconveniences of the last several years for several more. In the second place, he was not happy with the bioavailability studies the EPA was planning. Chris Weis, the Denver toxicologist, had announced that he was going to force-feed pigs with soil from the site and then slaughter them to see how much lead they had taken up. When Patti Clapper heard about that, she said "We do not have any pigs living in Smuggler—just people." Katherine Thalberg, who owns Explore Booksellers, a combination bookstore and coffeehouse that is one of the valley's landmarks, was so outraged by the proposed test that she wrote an angry letter and dispatched it to the local newspapers. It read, in part:

The proposal to imprison pigs in laboratories and stuff lead-laced dirt down their throats is unethical and macabre, and will undoubtedly yield totally useless results. One doesn't have to be a scientist to know that these conditions will cause illness to any creature whose normal diet is not dirt with or without the lead added. Nor do we have to perform experiments to know that lead is not edible. . . .

Pigs are extremely intelligent and sensitive animals, thought to be a good deal smarter than dogs. An adult pig is more intelligent and more capable than any human infant, or than a critically retarded human adult. Have we the right to conduct horrible experiments on other creatures just because they are not of our own species?

When I ran into Tom Dunlop a few days after the EPA announcement, he told me that he was in the process of designing tests that would show, once and for all, whether the lead at the base of Smuggler Mountain was in any way bioavailable. He said he hoped the EPA would cooperate in setting up and helping to pay for the tests, but he would seek other public funding as well.

Later, I asked Frank Peters why, in his view, the EPA had not simply delisted the site. He said that it was a matter of precedent. He told me the lawyer representing the EPA in Aspen had said to him "I've got twenty other cases going on now. If we let you be delisted because we haven't proved that there is a biological hazard here—because we haven't come up with a goat with five legs, or whatever else you need to have to prove that there is a problem here—then I am going to have to prove that in every other one of my cases. The EPA has never stood on that principle." The EPA, Peters told me, considers itself to be acting prophylactically. "Its mission is to go out and find hazards before they hurt people," he said. "Here we say we have had the hazard with us for a long time and apparently there is no ill effect. We think that that lack of evidence should be considered. And they say, 'We can't consider it. We have to look at the potential risk. We don't have to prove a present risk, just a potential risk.'"

While these events were unfolding, an enterprising reporter from the *Aspen Times* tracked down David Boon, whose 1981 master's thesis project started all the trouble. He discovered that Boon had become a professor and the co-chairman of the Hazardous Materials Management Program at the Front Range Community College, in Westminster, Colorado. Boon was forthright. "In my opinion, [the Aspen site] should never have been listed," he said. "I fought like crazy to get it off the list."

By an interesting coincidence, at just about the time that James Scherer was talking to the city council in Aspen, the director of the Superfund at the time, Henry L. Longest II, wrote a letter to *The New York Times* responding to an editorial entitled "In the Clutches of the Superfund Mess." Among the points that Longest made was one that caught my eye. He wrote "All 1,200 sites on the priority list have been screened for danger, and, when necessary, threats have been eliminated." It was a remarkable statement considering that the EPA had already spent an estimated $7 million—most of it in legal and administrative fees—in Aspen alone, and nothing had been cleaned up. I have no idea whether or not Mr. Longest has ever visited Aspen. If he does, he might like to take a stroll through the Smuggler Mobile Home Park. Some very nice people live there.

In midsummer of 1995, I took a walk with a friend up the dirt road that leads to the top of Smuggler Mountain. The road, occasionally driven by four-wheel-drive vehicles, is more commonly used as an exercise trail. At its lower end are the Mollie Gibson Park and the Smuggler Mine. At the top

is a nice viewing platform made of wood. It is about a 1,000-foot rise between the two, over a distance of a couple of miles. The young and hardy
run or mountain bike up the road. Those of us who are chronologically
challenged go up at a pace better suited to our age and station. Along the
road one gets wonderful views of the valley. One can see Aspen Mountain
and Red Mountain and the rest of the familiar landmarks. The Smuggler
Mobile Home Park, and the adjacent racquet club, are laid out like a low-
altitude aerial photo-map.

In any event, as my friend and I were walking along she looked
down on the area next to the racquet club. She noticed what seemed to be
a construction site. There were a few bulldozers as well as some trucks
and backhoes. Several men appeared to be busy moving dirt. She asked,
quite reasonably, "What are they building there?" I was tempted to answer "A monument to human folly," but I didn't, only because that would
not have distinguished it adequately from the many other construction
sites in Aspen where, for example, multimillion-dollar homes are being
built and then occupied for only a few weeks a year. Instead, I brought
her up-to-date as to what had happened with the Superfund since the
summer of 1991.

In February 1992, there was a meeting among the two Colorado Sena-
tors—Tim Wirth and Hank Brown—a delegation from Aspen and some
EPA officials. Perhaps the most significant thing to emerge from the meeting
was the formation of the Smuggler Mountain Technical Advisory Committee, made up of experts on lead toxicology and the like. The EPA and Aspen
agreed on its members—six in number, including Iain Thornton, Director of
the Global Environmental Research Centre at the Imperial College, who
came all the way from London. The committee met in October in Aspen and
released its final report—to which I will return shortly—at the end of January. Meanwhile, to assure that everyone's attention was suitably concentrated, the Justice Department, on behalf of the EPA, brought a $7.2 million

lawsuit against the Potentially Responsible Parties, which included Pitkin County and the Atlantic Richfield Company. An interesting feature of this suit was that each of the PRPs was sued separately for the full amount. This meant that if anyone dropped out, the others would still be responsible for the entire $7.2 million. But it also meant that if anyone settled for a share, this would benefit all the others. At some point, the Atlantic Richfield Company settled for something less than $2 million, and it is rumored that some of the other PRPs also settled. Keep in mind that none of this money went to cleaning up anything.

The final report of the Technical Advisory Committee is an impressive document. There are seven pages of citations to the literature. Among them is one to Chris Weis and colleagues for an experiment they carried out in 1991 on young swine—the experiment that caused such an uproar in Aspen when it was announced. Cutting through the scientific jargon, it is quite clear that the committee came to about the same conclusion as Patti Clapper—"We do not have any pigs living in the mobile home park—just people." The experiments were not very reliable and the animal models extremely dubious. On the other hand, the committee cited some very interesting results on children, which show that the presence of food in the gut may buffer the stomach against the absorption of lead. They noted that "If children in Aspen have better care [than, for example, inner city children] and receive meals on a regular basis, the timing of food in the stomach may prevent them from reaching the high levels of Pb [lead] absorption found in fasting human beings." Not only are poor inner-city children malnourished, but this malnourishment may substantially increase their chances of being poisoned by lead.

The basic charge to the committee was to answer three questions, which I will paraphrase. The first question was: Is there a clear and present danger to the health of the people who live close to the site? The second: If the answer to the first question was "no," is there a "reasonable probability of such a threat developing in the future?" And finally: What, if any-thing, should the community do about it? To the first question, the com-mittee unanimously answered "no." To the second, it answered that "The Committee unanimously agrees that there is a possibility [emphasis in the original] of a future threat, but the likelihood is small. If the demograph-ics, land use and environmental conditions remain essentially unchanged at the Site, we do not anticipate any future realistic health threat." The committee then made six recommendations for remediation, which were essentially those that the community had been proposing all along: moni-toring the blood of young children; capping a bit of the site where there were exposed mine tailings with fresh soil; having vegetable gardens "planted in raised beds with at least 12 inches of clean soil"; and finally, the intro-duction of various institutional controls to assure that the site remained safe.

At first, the EPA—in the person of Brian Pinkowski, the project man-ager for the site—appeared to reject the committee's report. Pinkowski ap-parently believed that the EPA's mission was to uphold some higher standard. The report, he said, was "detailed enough for physicians, [but] just not detailed enough for a public health agency." Shortly after Pinkowski's remarks, however, the EPA decided to accept the committee's recommendations—at least in theory. This still left the pending lawsuits. Clearly no one was going to clean up anything with an unresolved multi-million-dollar suit hanging over their heads. In other words, it was now 12 years since Boon had made his discovery, and not one spadeful of contaminated dirt had been removed.

That brings us to April 1995. On the 10th, it was announced that the legal actions had finally been settled. Among the elements in the 99-page decree was the release of Pitkin County as a participant in the $7 million lawsuit. It also made Tom Dunlop's office responsible for implementing the cleanup along the guidelines proposed by the committee.

The men my friend and I saw working on the site were doing so under Dunlop's direction. The cost of the remediation, Dunlop told me, will be around $400,000—roughly what the county was going to spend a decade earlier. "We have come full circle," he remarked.

While the citizens of Aspen were relieved, they were not joyous. As Mayor Bennett remarked, "There is no justification for the eleven-and-a-half years—the absurdity we've been through." But at least, as of early summer, something was actually happening on the site and people were beginning to think about the lessons. Certainly there is blame to go around. In retrospect, the county commissioners should never have invited the EPA into Aspen without first understanding the consequences. But a more enlightened agency, once having seen the absurdly small problem in Aspen—relative to the real problems with our environment—would have given local residents a pat on the back and told them to carry on along the lines they had in mind. Had this been done, the whole matter could have been dealt with for less than $500,000 in 1981. It is nonetheless important to note that no one I spoke to in Aspen would use this fiasco as an argument to abolish the EPA. The work of this agency is much appreciated. To give one of many examples, because of the high cost of living, most of the people who work in Aspen can't afford to live there. They commute, which has created a major traffic and pollution problem. If the EPA had not succeeded in establishing emissions standards for automobiles and lead-free standards for gasoline, the situation would be incomparably worse than it is. One wants the EPA to flourish, but also to make sense.

Language

In one of my many conversations with Hans Eberstark a few years ago, he told me that to his "eternal disgrace" he had not learned to speak Chinese "properly" during the nearly eight years he spent in Shanghai during the Second World War. He and his family had come to China as refugees from Austria. This is not to say that Eberstark was unable to carry on a conversation in Chinese. "I manage to muddle through," he told me. "The little I learned was Shanghai dialect. I think my pronunciation is pretty good. But my Mandarin. . . ." Eberstark emitted a sigh.

The operative word here is "properly." Eberstark's standards for speaking a language properly are different from most people's. During his professional life he was an interpreter. Until he took early retirement in 1987 he was permanently employed by the International Labor Organization (ILO) in Geneva. He then worked as a freelance interpreter and taught courses at the University of Geneva's École de Traduction et d'Interpretation, now more than a half-century old and regarded as the foremost institution of its kind in the world.

Eberstark and his interpreter-colleagues divide their working languages into two classes: A and B. The A—or "target"—languages are those that an interpreter is prepared to interpret into, and the B languages are those that he or she will interpret from. Eberstark's A languages are English and German. He mastered the former in Shanghai, where school classes were taught in English; the latter is his mother tongue. Eberstark's B languages form a continuum, along with his C, D, E, F and God knows what others. They include French, Dutch, Italian, Spanish and Catalan. "My Portuguese would be wobbly," he once remarked to me. "I wouldn't say Portuguese Portuguese, but Brazilian Portuguese if spoken slowly. Then another grade below would be the Scandinavian languages—Danish, Swedish and Norwegian. Then there is another language, which I forgot to mention because I hardly ever used it in my normal interpretation, but I do speak it, and fairly well. This is Surinamese-Creole, the language of Surinam. Also— but less so than Surinamese—are Haitian-Creole, and Papiamento, the language of the Netherlands Antilles."

There is also Yiddish, which Eberstark said he learned in Shanghai. When I remarked that that sounded like some sort of Jewish joke, Eberstark, who is very fond of jokes, said "Our whole life is a joke. God sometimes seems to have a sense of humor."

He went on: "I could manage to pass for a Swiss in Bern Deutsch [Barndutsch], in Basel Deutsch [Baseldytsch], though less so, and possibly in Zurich Deutsch [Zuritutsch]. I have done quite a bit of interpretation from Albanian; still I would say that my Albanian is flimsy, and I prefer not to use it except with friends. And Hebrew, to a certain extent, but I would not even interpret *from* Hebrew unless it's very elementary. I have had very few experiences of traveling as an illiterate. When I was in Ethiopia, I put together a basic vocabulary of Amharic, and I managed to make myself understood. I wouldn't say that I speak Amharic, but since it is related to Hebrew there were quite a number of similar word groups. I didn't have to go and learn it from scratch. That is one of my methods. I don't learn the words, I learn the differences. I anticipate what the words will be, and then I can look them up."

Eberstark's efforts to make himself understood have been made easier by his command of accents in most of his languages. He used to give lecture-demonstrations during which he would ask people in the audience—usually Swiss, who seem to have a different language in each valley—to converse with him in their native dialect. Before long, Eberstark would be conversing in the same dialect. This facility also applied to languages he knew virtually nothing of. "I certainly wouldn't claim Finnish as one of my languages," he told me, "but if I read a Finnish text to you, I could do it with a perfect Finnish accent, and a Finn who listened to me without seeing me [Eberstark has swarthy features and dark-gray frizzly hair, and looks very un-Scandinavian] would probably think I was a Finn. Let me add something. We had a teacher at the Oriental Institute in Vienna in 1947 whose name was Stefan Wurm. He is a little older than me—I suppose that he must be in his early seventies—and he spoke about thirty or forty languages. He also had an excellent pronunciation of most of them."

I have no notion of Professor Wurm's general intellectual abilities, but one of Eberstark's has been an incredible memory. He seemed to have a total recall of names, dates and events, and of numbers as well as words. Eberstark had a friend named Hank Wint who, until his retirement, was a computer specialist at CERN, the international elementary-particle physics laboratory outside Geneva. A few years ago, Wint learned from the *Guinness Book of Records* that there was a record for the number of digits of pi (3.1415926535. . . and so on) that anyone had committed to memory. The record at the time was about 3,000 digits. Wint bet Eberstark that Eberstark could not beat the record. As part of the bet, whatever number of digits he attempted to commit to memory had to be perfect. Eberstark lost. He had tried to memorize 6,000 digits and made a mistake at around 5,430. He was so annoyed with himself that he memorized 11,944 digits. This time he got all of them right and wrote them out—twice. It was Wint— whom I knew from my visits to CERN—who introduced me to Eberstark. After hearing this story I asked Eberstark how many digits of pi he still remembered. He said he wasn't sure, since he hadn't thought about it for a while, but he would see. He then rattled off 1,500 digits, and would have gone on—I think, indefinitely—if I hadn't stopped him.

I had been hearing about Eberstark for years from Wint and others, and I had also read about him in a fascinating book by Steven B. Smith entitled *The Great Mental Calculators*. Smith devotes a chapter to Eberstark, and Eberstark wrote the book's introduction, but I had no sense of what Eberstark would be like in person. The only other mental-calculating prodigy I had known was the late Wim Klein, who worked at CERN as a programmer. (Before the advent of electronic computers he did much of his computation in his head!) Klein was a very strange man indeed. Before he came to CERN in 1958, he had worked throughout Europe in circuses as a sort of calculating freak. I have an ineluctable memory of his phoning me at an ungodly hour of the morning to announce that my telephone number was a prime—a number divisible only by itself or one (17, for example). As early as it was, I had the presence of mind to ask him whether or not he had included the area code. Shortly thereafter, he phoned me back with the prime factorization of my entire number. Wim Klein could hardly sit still, and when he spoke he shifted rapidly from one language to another—his mother tongue was Dutch—speaking most of them haphazardly. He and Eberstark became close friends. In Smith's book, Klein is quoted as saying of Eberstark "He's great. He's absolutely daft, also he does not smoke. He does not drink. He's big and fat. He's so quiet. He never gets nervous. Just the opposite from me."

Hank Wint—who has now moved back to his native Holland—used to live in a large apartment not far from CERN. When I expressed interest in meeting Eberstark he said he would arrange a dinner for the three of us. He added that he had a Filipina cook named Gina, and that Eberstark—who is very fond of eating—greatly admired her Chinese-style cooking. This, he said, would probably induce him to make the trip from his home in Versoix, a small, lakeside suburb of Geneva.

When I arrived for dinner at Wint's place, Eberstark was already there. He was wearing a dark suit (I never saw him wear anything else) and a conservative tie. He turned out to be medium-sized and only slightly heavyset. Two things struck me: one was his expression, which had a kind of guilelessness about it—a look of childlike innocence. The other was his English, which he spoke with an undefinable accent. It sounded like the kind of English spoken by speech synthesizers, and it was uttered with an actorish clarity (an interpreter, after all, is a kind of actor).

The meal was excellent and Eberstark went through it with gusto. We had reached the egg roll stage when Eberstark said to the cook, "Gina, I think we will need you for a small experiment." Gina looked alarmed, but Eberstark assured her that the experiment would be painless. He asked whether her native tongue was Tagalog, the principal language of the Philippines, commenting that his own knowledge of it was almost nil. Gina's language turned out to be Ilocano, a variant spoken in the northern Philippines, and in fact, had been the mother tongue of the late President Marcos. Eberstark said he knew even less Ilocano than he did Tagalog. Then he asked us to pick 15 common words in English, and asked Gina to tell him what they were in both Tagalog and Ilocano. We picked words like "light," "apple," "school," "sunshine," "blue," "mother," "house"—15 in all. While this was going on, Eberstark was helping himself to more of the egg rolls. He took no notes, but would occasionally interrupt his consumption to make the odd comment. At one point, Gina got stuck on the Tagalog word for "sunshine," and Eberstark reminded her that it was *araw.* (Later, he told me "If you go to the Philippines, you hear the Tagalog for 'sunshine' often, because it is also the word for 'day.' It is aesthetically pleasing, so you make use of it in songs. It also sounds like the capital of the Swiss canton of Argau, and that, of course, would strike me.") Eberstark invited us to give him the 15 words in English, in any order. He then recited their Tagalog and Ilocano equivalents in what Gina said was a perfect accent.

I asked him which of the words, besides *araw*, he had known before. "I knew the word *ama*, which is both the Tagalog and Ilocano word for 'baby,' or 'child,' because the Indonesian word for child is *ana*," he said. "That was one of the first words I learned in Shanghai. When I went to the dentist there, he had magazines in Chinese, which I couldn't then read, but he also had magazines in Indonesian with a translation into English, which I *could* read. And I already knew the word *mata*, which is the word for 'eye' in both Ilocano and Tagalog, because the Indonesian word for 'eye' is *mata.* 'Day,' by the way, is *hari*, so *mata hari* is 'the eye of the day,' which was the nom de guerre of a well-known spy. *Hari* is related to *araw*, as you can see. *Mata* is also the word for 'eye' in modern Greek. Then there were the words 'light'—*ilaw* in Tagalog, and *silaw* in Ilocano. You will find that the word for 'light' in many languages has an *L* as a component—for example, as in *lux* and *lumen*. The moon in Hebrew is *lavana*, which means 'the white one,' and *lavan* is 'white, shining light.' The words for 'apple' and 'blue' were easy, because they are Spanish loan words—*manzanas* and *azul*, respectively. In Spanish, *manzana* is the singular and *manzanas* is the plural, while in Tagalog and Ilocano *manzanas* is both the singular and the plural. The name for this sort of exotic fruit would be taken from the colonizers as *manzanas*, since apples are bought in quantity. You wouldn't buy just one apple."

I asked Eberstark how long he would remember all that. He looked at me with a somewhat puzzled expression and replied "Always."

Eberstark was able to make use of yet another ability essential to his work as a simultaneous interpreter: He could listen and talk at the same time. (Simultaneous interpreters translate a phrase two to five seconds after it is spoken; consecutive interpreters translate at convenient breaks in the discourse.) At first, this may not appear to be a special ability—we all know people who seem to do it regularly—but just try it. As an exercise, I made use of the weather reports over the phone in Geneva, which are in French. I would call the weather number, and when the report began I would start translating aloud simultaneously. It would go smoothly for a few seconds, but then something about the weather would distract me, and another part of my mind would begin to think about that. Pretty soon, I would find myself getting behind and interpreting consecutively, which is much easier.

This kind of simultaneous translation was second nature for Eberstark. The night we met, Wint tested him by reading aloud a passage in Dutch while Eberstark rendered it into English. There was no discernible time lag. Eberstark explained this by saying he had a "three-track mind"; that is, he would listen to the Dutch, render the English, and think about the meaning of what was being said—all at the same time. The thinking is essential because the interpreter attempts to capture both the style of the speaker and the case the speaker is trying to make.

Like most gifts, simultaneous translation can be developed—which is what happens at the École de Traduction et d'Interpretation—but one is either born with it or one is not. Eberstark told me that many very good translators cannot do simultaneous interpretation because of the strain of keeping up. He has seen such people rush out of the translation booth screaming from the tension. I asked him if he had ever felt such tension, and he responded with an anecdote. A few years earlier, he said, he had worked at a conference in Germany involving occupational health services for German factory workers. One of his fellow interpreters was an inveterate pipe smoker who smoked while he worked. Listeners could hear various wheezing sounds through their earphones. One of the health workers became concerned and entered the booth to see if the man was having an asthma attack. When he saw that it was only pipe smoking, he began discussing the stress that he imagined interpreters must work under. Eberstark dropped by the booth as this conversation was unfolding, and a scientific test was proposed: the health worker would measure an interpreter's blood pressure before and just after he worked. Eberstark volunteered. His fellow interpreter warned the health worker that Eberstark might not be entirely typical. "If anything," he remarked, "interpreting calms Eberstark down." The health worker was exceedingly skeptical and the experiment was carried out.

"Afterward, I did have lower blood pressure," Eberstark told me. "But actually the pressure was very high both before and after interpreting. They advised me to take pills to reduce my pressure. I told them that the reason my pressure was so high was that I had eaten in their canteen. Normally I don't eat pork but that was all there was. It was very salty, so that probably was the reason. 'Why don't you test me tomorrow or the day after? Then you will get the real values,' I said, and, of course, I never went to that canteen again. When they retested, my blood pressure had returned to normal. That was a few years ago. Now I am older, so I do feel tension occasionally when I interpret. Put it this way: at that time, I never understood why interpreting was considered a stressful profession, and I thought that this was a professional lie meant to justify the fees that are paid to us."

I asked Eberstark if the fees were excessive. "They are high and not high," he said. "Not high, if you consider that ours is not a regular job. On the other hand if you look just at our fees, they seem high. We are now getting five hundred Swiss francs—about four hundred dollars—for a day, which is fairly high. If we work in a smaller team so that our hours are longer, it's something like eight hundred Swiss francs a day. But since meeting you last month, I have not had an interpreting job. My next one is next month, and I will not have another one before the end of the following month."

By the time Eberstark and I had this conversation, I had seen him several times a week for a couple of months. Most of the meetings took place in his attractive house in Versoix, which he shared with Leni, his German-born wife of 30 years, and two cats. (They have two grown daughters.) I was immensely curious about Eberstark's background and early life. When and how did these remarkable gifts manifest themselves? I soon learned that asking Eberstark about such matters carried a built-in risk. Because of his total recall, asking him even the simplest question was like trying to get a drink of water from a fire hydrant. If I wasn't careful I found myself retracing Eberstark's life in real time. Eberstark recognized the problem, and by imposing what he regarded as Draconian discipline, I was able to distill the essence into some 12 hours of tape.

The first thing I wanted to know was whether there had been a family history of this level of intellectual ability. Eberstark told me there had been noted Talmudic scholars on both sides of the family. On the paternal side, there had been a renowned commentator in the 16th or 17th century whose pen name was Mahazith ha-shekel ("half a shekel"); on the maternal side, there was an early-19th-century scholar who was noted for his piety. It appears that this ancestor, when he was very old, decided to learn to play the harp. "My death is approaching, and I wouldn't want to be an ignorant angel," he said. "As angels play the harp, I'd like to learn to play the harp."

Eberstark's father's family came from Bielsko-Biala in what is now Poland and was then part of the Austro-Hungarian Empire. His great-grandfather had no regular occupation, except as a *klezmer;* he played the fiddle at weddings and feasts in return for drinks, and he died on his way home from a wedding after falling asleep in the snow. His son, Eberstark's paternal grandfather, then age 13, took over the family responsibilities. He became a merchant and managed to send most of his brothers and sisters to school, after which they emigrated to America. He stayed behind and married a girl of 16 when he was in his late 30s. By this time he was the wealthiest man in Bielsko. A few years later, his wife—having given birth to two children, one of whom was Eberstark's father—decided she wanted to learn English, so Eberstark's grandfather found her a student-teacher who knew the language. The two fell in love, and without giving a reason, she told her husband that she would like to go live in Vienna for a while. He agreed and paid for her to stay there. Letters from her arrived regularly, but by the time they were sent, she was already in America with her boyfriend. She had written a number of letters, which she had given to a friend to mail at intervals.

Eberstark's father was supposed to go to America too. He had finished his *Gymnasium* studies in Bielsko in 1912, and as a reward he was given a ticket to the United States aboard the *Titanic*. But before he could go, he came down with a bad case of scarlet fever and had to give up the trip. When the First World War broke out he was drafted into the Austrian Army and served as a medical assistant. After the war he went to Vienna to study medicine. His sister, who was older than him, was by now running the family, and it was her idea that he should become a doctor. Eberstark commented "My father really didn't like medicine. He was more of an artistic type. He wrote excellent verse in German, and he was very good at drawing bacteria as seen through a microscope—blood cells, anatomic paintings. He was far more interested in doing portraits and writing verse, and card playing in the coffeehouses of Vienna than in his medical practice." It was, however, through his medical practice that he met Eberstark's mother. She was a patient in a sanatorium where he was working as an intern. She had been married before and she was poor, neither of which sat well with Eberstark's aunt, who had hoped that her brother would find a rich woman among his patients in the sanatorium. But love won out and the two were married. Eberstark was their only child.

Eberstark is quite sure that by age two-and-a-half he was able to read. He recalled a none-too-pleasant event that occurred on his third birthday. A friend of his mother had given him a book of fairy tales specifically because she doubted that he could read, as she had been told. Eberstark took the book and began scanning it with great interest. The woman insisted that he read aloud. He recalls being very rude to her because he wanted to get on with his reading. Ultimately he was persuaded to read a few sentences from the middle of the book, but the woman suspected some sort of trick (a common reaction to Eberstark's feats of mental gymnastics), and she insisted that Eberstark read aloud from the end of the book. He refused because he wanted to read the story in its proper order. His parents promised him that if he would read a few sentences from the end they would all leave him alone. He did so and the surprised woman said "But he can read!" "I could have told you so in the first place," Eberstark retorted.

At the age of six Eberstark began studying Hebrew after school. Then he found a conversational dictionary—a *Konversatsionslexicon*—that included the Greek and Russian alphabets. Eberstark remembers his early education as something of a problem. "My parents didn't tell my teachers that they should prepare books for me because I was already able to read, so I was in these reading lessons, and since the other children stammered and stuttered while trying to read, I did the same thing, because I thought that this was the done thing in this school. Reading lessons were extremely boring for me. I received the reader at the beginning of the year, and on the way home—we lived in a suburb of Vienna, so it took thirty-five or forty minutes to come home—I read two-thirds of the book. I read Shakespeare in German at age six or seven. I liked it, but I don't think I understood it all."

Until the Anschluss—the union of Austria and Germany in 1938—Eberstark went to another school where the standards were high and where a number of students were interested in reading and other intellectual pursuits. But deteriorating social conditions forced the family to move into the Jewish district of Vienna. "My first school there was pretty bad," he recalled. "There were always attacks by the Hitler Youth, who beat up teachers and pupils. This did not happen to me due to what seemed to have been my mother's extrasensory perception. She would say, 'You can go to school now, but not the day after tomorrow. Stay home for two days and then you can go back.' When I went back, my classmates told me that they had been raided by the Hitler Youth. I never witnessed this because of my mother's foreknowledge. Then I was transferred to another school, my last in Vienna, which had a tacit agreement with the Hitler Youth that it would not be raided. "Life at this school was a real pleasure," Eberstark said, explaining "It was particularly stimulating because the pupils were treated as an elite. The teacher said, 'We don't know where you will go, or whether you will survive, but we will teach you things you aren't expected to know at your age. We hope they will serve you well.' That was the cream of the cream of the pupils. Almost none of them survived."

Eberstark himself barely survived. In desperation, his parents decided to have him adopted by a Dutch family; one of the families that offered their services was a branch of the DeBeers diamond family. Neither they, nor another that also offered to adopt him, survived the war.

The circumstances that enabled Eberstark and his family to get to Shanghai almost seems like another Jewish joke. His father had obtained visas for Trinidad, but before the family could sail, the British stopped a migration to their colonies. Just as his father was returning the tickets to the travel agency, another man came in to return two first-class boat tickets to Shanghai. "My father said 'I'll take these,' and he managed to get another one for me," Eberstark recalled. "My parents spent their last money to get those tickets to Shanghai, and visas. We were among the very few to have visas for China. If they hadn't spent that money, it would have been confiscated anyway."

The Eberstarks traveled by way of Switzerland to Trieste, where on February 8, 1939, essentially penniless (they had been allowed to take out a total of 90 reichsmarks), they sailed as first-class passengers for Shanghai on the *Conte Rosso*. The trip lasted 25 days and Eberstark remembered it as idyllic. Despite the fact that the menu had seven or eight pages, he ate the same thing every night: hors d'oeuvres, spaghetti with meat sauce, cherry compote and a double helping of ice cream. He also began learning Italian. He vividly recalls a stop in Colombo: "We were surrounded by very picturesque crowds in colored clothes, naked children, and so on. Everyone tried to sell us souvenirs, but we had no money, and when they couldn't sell us the souvenirs, they gave them to us. Wherever we went, the poor people invited us in, because they knew that we were even poorer than they were. I must say that this was quite a contrast with the treatment we had had from the Europeans—not just Germans and Austrians, but Europeans. The British and the French were responsible for closing their colonies to us and leaving us to our fate."

Upon reaching Shanghai, the refugees were loaded onto cattle trucks and shipped to a camp located in the northern part of the city. Shanghai had been occupied by the Japanese since 1937, and a great part of the city was in ruins. The camp itself was sponsored by Jews who had emigrated to China earlier in the 1930s and prospered there. Eventually, there was a community of some 19,000 Jewish refugees in Shanghai, most of them from Russia and Poland. The lingua franca of the camps was Yiddish, which Eberstark readily mastered. At first, the refugees were segregated by sex, and Eberstark and his father lived in a dormitory with 40 other men. There was, however, a pressing need for doctors, and Eberstark's father volunteered; in return, the family was given a room, which it shared with another Austrian family. "My father got a small amount of money as well as our board and lodging for his work," Eberstark told me. "A few months later, he was promoted to medical superintendent and chief physician of another refugee camp, so we moved there, and then we had a mini-apartment, so in a way we were better off than most of the others. At first the food was fantastic because it was prepared by Chinese cooks. There was a sort of sweet porridge that I liked very much but which looked disgusting, so most people were put off. As a result, the Chinese cooks were dismissed, and some Austrian refugee ladies took over the cooking, and the quality of the food went down. But everybody else was satisfied because it met their demands. I wasn't, but then who was I at the age of ten?"

As Eberstark recalls, the Jewish refugees and the Japanese, who occupied Shanghai, lived in relative harmony until Pearl Harbor. This was partly due to the presence of a German consul in Shanghai who took the view that Austrian refugees with German passports—albeit stamped with the letter "J," something the Swiss had done—were in fact German citizens and therefore entitled to consular protection. This remarkable individual did not last very long and was replaced by a more conventional German consul who declared that the Jews were stateless and thus not entitled to any protection.

After Pearl Harbor, the Japanese decided that since the Jews were enemies—so it seemed—of their German allies, they should be treated as such. An order was issued requiring all Jews to live in a specially designated area of the city. Anyone without a residential permit would be jailed, which was, in effect, a death sentence, since the jails were infested with typhus-carrying lice. To leave the ghetto required a special pass. Eberstark remembers with some distaste an "absolutely crazy" Japanese official who was in charge of passes. "You might stand in line for hours for a pass and then, for no reason, get a beating." In 1945, near the end of the war (although they had no way of knowing that), Eberstark's father and the other doctors managed to pay back the Japanese in kind. By this time, the Allies had begun bombing Shanghai in earnest, and the Japanese—unbeknown to the refugees— had moved huge armaments factories, spare-parts depots and the like into the areas where the refugees had been required to live. It appears as though the Allies had been aware of this and held off bombing that part of Shanghai, but in 1945 that began to change. The Jewish physicians were treating not only their own bomb casualties but also those of the Chinese living in the designated areas. On July 17th, a major citywide air raid resulted in many Japanese casualties. The Japanese authorities called on the Jewish doctors to leave the ghetto and treat their wounded as well. The doctors replied, "No, we can't, because we don't have special passes to leave the ghetto, and we don't want to violate your rules." Eberstark is convinced that if the war had not stopped soon after, most of the refugees would have been killed, either in bombing raids or by the Japanese.

Almost immediately after reaching Shanghai, Eberstark entered the Shanghai Jewish Youth Association school, which had been set up by the refugees but financed by a Mr. Kadoorie, an Iraqi-born Jew. Almost all the teachers were German-speaking, but the classes were in English. Eventually, Chinese and Japanese were introduced as well. Then the Allied air raids began, and there were blackouts every night for nearly two years. "It was practically impossible to read at night, which was very depressing," Eberstark recalled. "Then the war was over. You cannot imagine the jubilation. First the Nationalist Chinese—the Army of Chiang Kai-shek—came in, and then the Americans and some British soldiers as occupation forces to help the Chinese. There were lots of jobs and fabulous salaries—something like eighty or a hundred dollars a month. Totally incredible, because before, the average monthly income had been five or ten dollars. We got our German or Austrian passports back, and by 1946 there were several places we could go. The bulk of us went to the United States. Some went to Canada. Quite a lot went to Australia. Quite a lot more went to Israel. A very few went to South Africa. Some people—either because of nostalgia or because they found the idea of going elsewhere more difficult or because they hoped for reimbursement or compensation—went back to Austria or Germany. My father was in that category. He also thought that I should go to the University of Vienna. Personally, I was not enthusiastic about going back at all. For one thing, I felt at home in Shanghai, and I thought that I would feel very much like a stranger in Austria. But at that point I was eighteen, and so I felt that it would be difficult to stay in Shanghai on my own."

The Eberstarks landed in Naples on February 8, 1947—eight years to the day they had left Italy for Shanghai. After crossing on the *Marine Falcon* they traveled from Naples to Vienna by cattle car. For a while, it looked as if Eberstark would have to stay in China after all. He had received his "leaving" certificate from the Shanghai Jewish Youth Association school, and needed his examination scores to enter the university. But the examinations were to be given after he was scheduled to leave China with his parents. "My mother was hysterical to find that at the end of the war we would be separated and she might never see me again," he said. "But we made an arrangement, and I was able to take the examinations and still get to the boat on time." After their five-day trip by cattle car to Vienna, "We were received with open arms by the mayor of Vienna, who was certainly not a Nazi during the war but a socialist general," he recalled. "That was one of the shocks—the pleasant shocks—I experienced. I thought it was going to be enemy country. I insisted on wearing a Star of David—a Magen David—because I did not want to be taken for an Austrian. I wanted to show my nationality. As a result, I never came up against any anti-Semitism, because I made it quite clear that I was Jewish."

Eberstark's father, who resumed his medical practice, decided that Eberstark should become a chemist. This led to violent arguments, since Eberstark had long before decided that he was going to study languages. His father told him that he could never earn a living as a linguist. Eberstark remembers him shouting that there were people who spoke five or six languages and were poor. "They go *neben die Schiech* [an Austrian Yiddish expression meaning 'beside their shoes']. Chemistry is the science of the future, and you'll be part of it. Why don't you study chemistry? You're interested in seeing liquids turn yellow and green and foam, and the rest." Eberstark countered by saying that mixing liquids was dangerous and that he might easily have an accident that would finish him. Besides, if he became a second-rate chemist he would have to work "someplace in lower Austria." "Going to the country," he added, "was always terrible to me, because I have been raised in towns and have never had any predilection of greeneries."

In the end, despite the dire parental warnings about poverty, Eberstark enrolled in linguistics at the University of Vienna with the intention of taking a doctorate and becoming an interpreter. He had been told that the entrance examination for the interpreters' school was very hard and that most people failed it. Having never had the slightest difficulty with any school examination, he decided to take it. The examination was in German and English; in the course of it, the examiner asked Eberstark what his father did, and when Eberstark replied that he was a doctor, the examiner asked him to name several diseases in English. When he mentioned typhus, he was asked for the German and was able to tell the examiner that it was *Flecktyphus,* typhus being the German word for typhoid fever. The examination, it appears, was full of traps like this, but he had no trouble passing it.

When Eberstark entered the university, simultaneous interpretation was being used for the first time, anywhere, at the Nuremberg trials. It was so new that it was not yet being taught at the Vienna school. Consecutive interpretation was still the standard. If Eberstark had stayed in school for just the conventional four years, he would have missed learning simultaneous interpretation, which was introduced into the curriculum around 1951. As it was, he became almost a perpetual student, attending university classes for some 15 semesters—eight years—and finally emerging with his Ph.D. in January 1956. Once classes in simultaneous interpretation were introduced, they were taught by using telephones. A student outside the classroom would listen while a text in either English or German was read to him or her—a sizable proportion of the students were women— by another student inside the classroom; then the situation would be reversed. "This was totally inadequate," Eberstark told me. "But these were pioneering days. If you mastered this kind of telephone translation, then the work with modern equipment was that much easier."

Eberstark had planned to write his Ph.D. thesis on sound symbolism—the way the meanings of words like "whir" and "buzz" are contained in their sound—and its relation to semantics, with special emphasis on English. But the plan was derailed by the Bible— or, more exactly, by the Bibles. "I had a fantastic number of Bibles," Eberstark explained. "Not because of any special theological interest but because the Bible has been translated into the largest number of languages possible—some twelve hundred or so. Also, the Bible was translated by translators who really put their hearts into it and—provided that they were not too awed by what they thought was the original text—dared to use the idioms of the target language. Often, as with some of the African languages, the Bible was the only literature available.

"There was one language that fascinated me. It was called Nengre Tongo—Negro Tongue. Today it's called Sranan Tongo or Surinamese—Surinam Creole. There had been some work done on it, but this was on the language spoken in the bush, which is quite different from the language spoken in big towns like Paramaribo. It is in a way a perfect language. There are no redundancies in grammar. Most languages suffer from redundancies in grammar. German is a classic example. If you abbreviated every word, you could still read the sentences, because most of the inflectional endings are redundant.

"In all the creolized languages—including, to a certain extent, English—the grammatical redundancies have been worn away as a result of the contact between two peoples who had to begin communicating with each other by means of a pidgin language. Creole languages normally grew out of pidgin languages, and pidgin languages are not the languages of anyone but are used as languages of intercourse—as vernaculars only. Creolized languages develop a life of their own, and acquire far more synonyms, allusions, proverbs and so on, while keeping the grammar of the pidgins; namely the minimum. Surinamese attracted me because there was nothing of the highfalutin' in it."

Warming to the subject, Eberstark went on, "I began to learn Surinamese using the text of the Gospel of Saint Matthew in English, German, and Dutch, and comparing them to the Surinamese version. I wrote down rules for Surinamese and tried to use them to make my own translation—to see where it differed from the official version.

Where the King James version said, for example, 'so-and-so begat so-and-so' and 'Mary became pregnant,' it said in the Surinamese Bible that 'so-and-so made so-and-so' and 'Mary got a belly'—*kissi bele*.

"Actually, the Surinamese Bible I used had not been translated by native Surinamese but by various missionaries who had not really mastered the language. There was no feedback from their parishioners, because they took what they were taught as the gospel truth. You find the same situation with the King James Bible, which is written in very bad English, although it does reproduce Semitic phraseology. I tried to analyze the language and show that it consisted of different layers. The oldest layer was the pidgin, which was used when the slaves were transported from Africa to Surinam. These words were of Spanish or Portuguese origin. There were very few of them; the main infrastructure was English. But the English had been changed considerably when it was adopted into pidgin. There was a superstructure of Dutch because in 1815, the Dutch took Surinam over from the British. I had never met a real Surinamer, but I derived what I thought Surinamese sounded like from what I learned from the Bible translation. Later on I got the chance to speak it."

While Eberstark was working on his thesis, he became known as the computing prodigy of Vienna. He had been an outstanding student in all sorts of arithmetic, though the subject was of no particular interest to him. In 1951, however, he happened to read some articles in a Viennese newspaper about two mental calculators: one was a college student in Milan; the other a young Indian woman named Shakuntala Devi, who was giving exhibitions while touring Europe with an impresario. The college student extracted roots of large numbers in his head. Eberstark was also able to do this, and he wrote to the newspaper saying that he knew the secret, hoping to get a free subscription.

He got more than he bargained for. He was interviewed by one of the paper's reporters who told him that Shakuntala Devi was coming to Vienna for an exhibition and that Eberstark ought to serve as her local counterpart. At this point Eberstark found out what the young woman actually claimed to be able to do. Apart from multiplying six-digit numbers in her head, she could, he was told, find the day of the week corresponding to any date in history. Eberstark recalls telling the reporter that he could not do any of those things, and being told not to worry as long as he could extract roots. He was then taken by someone from the newspaper to the Indian Consulate, which was organizing Shakuntala Devi's tour.

"At first," Eberstark told me, "we were received politely. They showed me some press clippings, and I was suitably awed. Then the Indian consul told me that wherever Shakuntala Devi went in Europe there were always people who said that they could do the same thing. 'I am sorry to see that you are among them,' he said. 'You are, no doubt, a very good calculator, but this is something completely different. This woman is an absolute miracle. She sees the figures before her eyes and she *reads them off!* No one knows how she does it; least of all herself.' It was proposed by the newspaper reporter that Shakuntala Devi and I pose for a photograph together. The consul said 'Certainly not! Clear out because I have important things to do.'"

Eberstark decided then and there that he would go into deep training to be able to do whatever Shakuntala Devi could do. When she came to Vienna he went to her performance, which seems to have been a mixed bag. For dates prior to 1900, her days of the week were off by one day—something the older audience members spotted, since she got their birthdays wrong. Her impresario explained to the audience that the error might be due to the difference between the Indian and the European calendar. He was not far off. She had not realized that 1900 was not a leap year; only centuries exactly divisible by 400—like the year 2000—are. She was able to extract roots, but when she was asked to multiply six-digit numbers as advertised, she said she was out of practice. Eberstark had by now mastered this skill and, having been sufficiently annoyed by his treatment at the Indian Consulate, he decided to give his own performance. He did this at the Austrian College Society, reproducing all of what Shakuntala Devi claimed to do—and more.

In preparing for it, Eberstark created what one might describe as a number script. He translated numbers into "words" in a language that he invented. Our minds seem to be more adept at retaining lists of abstractions if the abstractions sound like words. (To illustrate this point, Eberstark presented me with 20 letters, which he said I would remember. He recited b-a-u-t-c-h-p-i-r-g-n-a-r-g-d-a-l-e-p-o. My first reaction was a kind of panic, but then it dawned on me to group batches of letters into "words" like "bautch" and "dalepo," and I had no trouble remembering them.) Eberstark assigned a letter to each digit. For example, seven became L, because it looked like an upside-down L; likewise, six became P or B. Had he left it at that, he would have ended up with strings of unpronounceable words. To keep this from happening, he added a short A, which has no numerical significance. There is also *advanced* Eberstark in which 66 is not PP but M. God knows why. Eberstark remarked "I remember numbers because their shapes remind me of letters. I would call it not mnemotechnics, but mnemonics. I automatically see each figure as something written in a particular script like, say, the Greek or Cyrillic alphabet. I don't think of *anthropos* in its English transcription but as written down in the Greek alphabet. My number system was simply like learning a different script."

What has fascinated Eberstark about mental arithmetic is not the numbers—he has no interest in mathematics—but the odd sounds that emerge. The number 2,694 is "turf," while 2,693 is "treg." In advanced Eberstark, 504,562 is "sofspat." All of his mental arithmetic is done in terms of these "words." He recited the first 1,500 digits of pi for me and it sounded like a mixture of Tibetan chant and rap song. Eberstark told me that it took him about five or six minutes to memorize 100 digits. The difficulty, he explained—and this is how he lost his bet with Wint—was to glue together the 100-digit units. To do this, he had to find additional word associations.

The world of computing prodigies is a small one. So it was natural for Eberstark and Klein to be in correspondence before they ever met in the early 1960s; by then Eberstark had gone to the ILO and Klein to CERN. This was also about the time I first met Klein. I used to spend summers at the laboratory and Klein's office was not far from mine. He was a very friendly man, always willing to extract a root if one needed it. It turned out that he had memorized the entire table of logarithms to many places—essential for root extraction—along with the values of all the trigonometric functions. It saddens me to think that everything Wim Klein could do in his head can now be done more accurately—and in a fraction of the time—with a pocket calculator costing a few dollars. With the arrival of electronic computers at CERN there was not much for Klein to do, although he often gave lecture-demonstrations, some of them with Eberstark.

Eberstark told me about his relationship with Klein. "We were very friendly because he said that I reminded him of his older brother." (Klein was actually some 15 years older than Eberstark.) "His older brother [also a computing prodigy] had been gassed by the Germans. Wim himself was almost picked up by the Nazis in Amsterdam. He just managed to avoid the place where Jews were grabbed off the streets. [In some of the conversations I had with Klein he would begin telling me about the war and then break off as if he had seen a demon.] He also said that I had similar methods to his in arithmetic—and similar tics. But in some ways we were very different. As you know, Wim was impatient, choleric, and he often exploded. I'm very calm. But we had this common interest in calculation, although I was also interested in language—etymologies and so on—and he was not. Wim also liked to drink wine and alcohol in general, and I am closer to the Jewish tradition of temperance. It was difficult to communicate with Wim, because he bubbled over. But I knew what he wanted to say."

Indeed, when Klein calculated, he chattered along in several languages; one had the feeling that the mechanism was exposed, like one of those Swiss watches that are open in the back. When I watched Eberstark compute, nothing was visible; the answer came out, with no warning, all at once.

"I can't really compare myself to him," Eberstark continued, "because he was way ahead of me—apart from my memory for figures. For him, that kind of thing was a sideline, but it was the meat of the subject as far as I was concerned. It took us about the same amount of time to calculate the weekdays for a particular date. I was pretty fast at multiplication, but he was even faster. He wrote his multiplications as he went along, while I built up the multiplications in my head and only wrote the answer. It took him about two-thirds of the time that it took me. The one great thing about him was that he was able to factorize seven- or eight-digit numbers [like my phone number]. But he was unable to explain how he did it. He tried to explain but he lost patience. I could understand what explanation he managed—but to do that, you had to have some inside knowledge.* He was not a

*Klein also tried to explain the "trick" to me. One element was the introduction of remarkable arithmetic identities that he seemed able to generate spontaneously as needed. Where these identities came from he could not explain, except to say they were "obvious."

mathematician; he was a calculator. There's a big difference. Like mathematicians, calculators can work on the theory of numbers [Klein did.], but numbers themselves are of no particular concern to mathematicians."

After retiring from CERN, Klein moved back to Amsterdam. In July 1986, he was murdered by two drug addicts, whom he surprised in the act of searching his apartment for money. They were caught and later released for lack of evidence. Presumably they are still at large. When I commented to Eberstark how sad this seemed, he answered "Yes it was a sad ending, yet it was in character. He wasn't the sort to die in bed after an illness. There was something sensational, a certain circus atmosphere about him, and this was the crowning end to his life."

Eberstark must surely have one of the highest I.Q.s ever measured. I have certainly encountered no one like him. For that reason his commitment to astrology is beyond my comprehension. When he told me about it I told him that, in all candor, my attitude toward astrology was well-expressed by the great 20th-century German mathematician David Hilbert. Hilbert once remarked that if the ten smartest people in the world were put together in a room and told to invent the stupidest thing they could think of, they would not come up with anything as stupid as astrology. Eberstark responded by asking if Hilbert knew how to draw up horoscope charts, and I saw that it was useless to press the point. Needless to say, Ebenstark knows the precise time of his birth: January 27, 1929, at 15 minutes and 40 seconds past nine in the evening. He knows the precise time of his two daughters' births, since he was there with a watch. He knows that his ascendant—whatever that is—is 27 degrees and 37 minutes. He knows his current wife's horoscope—it was one of the reasons he married her.

The first time he married—at age 24—it was without the benefit of astrology. A friend of his had answered a marriage ad in a Viennese newspaper, but nothing had come of it. The friend kept the young lady's picture, however, and when Eberstark saw it he decided to marry her himself. She, in turn, wanted to marry someone with a university background, and by this time Eberstark was the computing prodigy of Vienna. He remembers the brief tenure of his first marriage with great nostalgia. "I look back on those days as paradise," he told me. The young lady felt otherwise, and after deciding that Eberstark was never going to stop being a student, she left him. "She wanted someone who would have a career with an international organization in Geneva and would travel a great deal," Eberstark told me, which was precisely what he went on to do.

Eberstark attributes his current marriage to a combination of astrology and Wim Klein. By the time he met Leni (then Leni Mergenthaler) in the fall of 1962, he had been working as an interpreter for 12 years and had moved to Geneva to work for the ILO. There was a German club in the city then, which was open primarily to German citizens. As an Austrian, Eberstark was made an honorary member, and he gave lectures at the club in both etymology and astrology. In the course of the latter, he would draw up simple horoscopes for various audience members and use them to divine their subjects' principal characteristics. A young woman at one of these readings was born on May 16, 1939, and Eberstark found this an extremely auspicious date. "I was quite surprised," he told me. "A comparison of my chart with hers indicated very many harmonious aspects. According to our horoscopes she would have been interested in me, and even been sexually attracted to me. Unfortunately, I was not attracted to her at all. I did her chart and forgot about her, except for pitying myself for having found someone who was attracted to me when there was no spark in return." So much for horoscopes!

Eberstark sighed at this unhappy recollection. "Two weeks later, my father came for a visit," he continued. "I was to give a performance at the German club. The room was full. The only two seats left were the ones that Wim and I had vacated at my father's table. Two girls came in late and asked whether those seats were free and if they could have them. So they sat at my father's table. One of the girls struck my eye as being very beautiful. When Wim and I went through the calendar business—doing the weekdays—I asked her what her birthday was. It turned out to be May 16, 1939—identical to that of the girl I hadn't been attracted to. So I had the essentials of her horoscope even before I knew her. That was Leni, and the other girl who was with her was her sister.

"Well, my father invited them back to our place, and the next time I saw her I proposed to her. She didn't take me seriously. She discouraged me from the very start, saying 'You're far too clever for me. You're not my type. I'm not interested in you except for the fact that I can learn a lot from you.' Then she went off to London. 'Don't expect me to come back,' she said. The only reason that I didn't give up was that I saw, by this comparison of charts, that she was meant for me. I did what was called a composite horoscope—fitting the midpoint of a planet on one chart with the same planet on the other chart—and I was absolutely sure that what I had was a typical marriage horoscope. And in fact we did get married. It took far too long, from my point of view. I met her on October 16, 1962, and we got married on August 3, 1965." It was in the stars.

After Wim Klein left Geneva, Eberstark's lectures focused on his conception of the origin of languages. His view, which he has held practically since his university days, is that at some epoch in the past there was one protolanguage, which was spoken by all our articulate ancestors—the first human language. Academic linguists used to dismiss the idea, but it now seems to have wider support. Eberstark thinks that many academic linguists are too narrowly focused. "They specialize in a given group of languages and hardly look beyond their noses," he said. "A person perfectly conversant with Indo-European languages would have had a classical education in Latin and Greek. He would know the Romance languages and understand the sound laws. But he would find the Finno-Ugric languages—Finnish, Estonian and Lappish on the one hand, and Hungarian on the other—to be completely different. He wouldn't see that the inflection syllables, the conjugations, are similar in Finnish and Hungarian, and that these in turn are similar to the Indo-European languages. The same correspondence can be found between the Semitic-Hamitic, the Indo-European languages and the Altaic languages like Mongolian and Japanese. People who are really interested in learning about the origins of languages, and try to analyze them outside the realm of Indo-European languages, are themselves often outsiders. As outsiders, they are rather undisciplined in establishing rules. The theory was initially discredited because many of the examples given didn't stand up."

Eberstark described some of the pitfalls. Loan words are often cited incorrectly as evidence of linguistic similarities. In Hebrew, for example, the word for "organization" is *irgun*, derived from the same root as "organization," which is from the Indo-European consonant root *'r-g-n*. Then there is the matter of phonemes—sounds that occur in a given language; 15 in Polynesian and some 45 in Georgian, to take two examples. The fact that similar phonemes occur in various languages, opponents of the unified-language theory argue, has to do with sound symbolism. A word like "roar" might sound similar in various unrelated languages because it sounds like what it means. To avoid this cavil, Eberstark chooses words that are primitive—verbs like "eat" and "drink" and simple numbers. He gave me a brief lecture on the words for "one":

"What you find in the Asian languages is one of the most commonly spread roots for 'one'—*whd* or *yhd*. It is *echad* in Hebrew, *wahid* in Arabic. Fine, but in Finnish it's *yksi*, which doesn't sound much like *whd*. But the Finnish word has been modified—*yksi* is the nominative, but the root is *yht* or *yhde*, which leads again back to *yhd*. Then the Finnish *yksi* becomes *e'gy* in Hungarian, but you have to know the phonetic correspondences for the Finno-Ugric languages to see how that works. Japanese is a special case, because 'one' has two different roots: the Sino-Japanese root, in which the Japanese took the old Chinese numbers, and the original Japanese numbers which are very different. In the original Japanese, 'one' is *hitotsu*, which when pronounced sounds very much like *yhd*. The usual Japanese for 'one' is *ichi*, and this is derived from the Chinese *yit*, so there again you have *yhd*. Then, in Malay or Indonesian it's *satu*. That doesn't sound very much like *yhd*, but in Malay there is no 'hy'—pronounced as it is in 'huge'—phoneme, which means that *satu* would have evolved from *hyatu*. There are a number of languages where the 'se' sound corresponds to 'yh' in another language—*sind* and *hindi*, for example, both meaning 'India.' In English you won't find *yhd* or *whd* meaning 'one.' But in Anglo-Saxon you do find *wiht* meaning 'something.' That is clear from 'naught,' meaning 'not something,' which was originally *nawiht*. In German you have it in *nicht*—'not'—nothing. You have it in *wicht* too. In German, *wicht* means 'little dwarf'; that is, a little something.

"There are also an enormous number of languages that have the same root for 'seven': *seben, septum, hepta*—the 'se' and the 'he' phonemes are related to each other—and *seitzma* in Finnish or *he't* in

Hungarian. In Hebrew it's *siva* and in Arabic it's *sib'a*. In Basque you have *saspiak*. In Swahili, it's *saba,* but the problem is that several of the Swahili numbers were borrowed from the Arabic, so *saba* is not really an argument. In Albanian you have. . . . " At this point I stopped Eberstark because my head was beginning to reel.

The École de Traduction et d'Interpretation [ETI] occupies the four upper floors of a modern building in downtown Geneva. The institution, which became affiliated with the University of Geneva in 1972, was founded in 1941 by a sort of visionary named Antoine Velleman, a language professor at the university. It was wartime, but he foresaw the need for interpreters once the war was over. In fact, most of the interpreters at the Nuremberg trials were trained at the ETI. By 1961, there were about 60 instructors who taught interpretation and translation in some 20 languages. But in 1968, the school realized that there were simply too many languages being taught, and the number was cut down to seven: German, English, Arabic, Spanish, French, Italian and Russian. A student who enters the ETI must have mastered at least three of these languages, French being required. There is a very stiff entrance examination, which has about a 95 percent failure rate. Applicants are given two chances. The interpretation course lasts four years, and it takes an additional 18 months of apprenticeship to get an interpreter's license. The ETI maintains a very small permanent faculty for its 320 students. Most of the faculty consists of people like Eberstark, who teach courses but continue to practice, thus maintaining contact with the profession.

When I was in Geneva, Eberstark was teaching a course in consecutive interpretation. He invited me to sit in as a sort of resource. I was to give a lecture that the students were to translate into German—the target language—as I went along. I had a prepared text, but it was not given to the students. This put them in the most difficult situation for consecutive interpretation. Usually, consecutive interpreters insist on having a prepared text in front of them—if one exists. By contrast, dialogue is relatively easy to interpret because of all the natural pauses in spontaneous speech. Eberstark told me that a prepared text had once saved him during a conference in Romania. He arrived in the country with no real knowledge of the language, but after four days he was able to read the local newspapers. On the fifth day, he was in the German booth when the order of the conference speeches was switched. The Romanian–German interpreter was nowhere to be found. Eberstark took the prepared Romanian text and translated it into German as the speaker went along.

When I arrived at the ETI at the appointed hour, I found Eberstark and three students waiting for me in a classroom. There were two young women—both German-speaking—and an Englishman who had taken a Ph.D. in political science at a German university. The classroom was fitted out like a conference room at an international meeting. The walls were lined with glass booths, where student interpreters sat. It looked very much like the real thing. I sat on a raised platform while the students sat at desks below me. Not quite knowing how to proceed, I began reading my lecture. Very soon, the students asked me to slow down; it is easy to speed up when one is reading aloud. As I continued, the students made notes. Eberstark had explained to me earlier that consecutive interpreters do not use a conventional shorthand. They do not take down literally what one is saying, which would be too slow and too cumbersome to read back. Instead, they invent an individualized mnemonic structure that enables them to reconstruct what was said more or less instantaneously. A typical interval for this is three or four minutes, but they must have the ability to reconstruct the text even if the speaker goes on for half an hour. After about three minutes Eberstark stopped me.

He called on one of the students, who then translated what I had said into German. It seemed faultless. But after complimenting her, Eberstark got down to what he called "the nit-picking." She had left out a few details and was stopped by some puns, which were probably untranslatable.

As I lectured, I was able to peek at Eberstark's own mnemonics. They were a fantastic mixture of words in different languages, along with what looked like Hebrew, Chinese and Arabic letters. From this he had been able to reconstruct—with no apparent effort—both what I had said and what the student had left out. What he—and Klein for that matter—can do, cannot really be taught. In thinking about it, I was reminded of a definition of genius I once heard: A genius is someone who can do easily what the rest of us cannot do at all.

———

This profile appeared in a substantially truncated form in the October 1993 issue of The Atlantic. *Some time afterward, a woman with a German accent called me and identified herself as a cousin of Eberstark's. She said she had just come back from Geneva and had seen Eberstark, who was very pleased with the article—although he thought it was too short. About a year went by, during which Eberstark and I exchanged the odd note. But then she called again—this time with very bad news. Eberstark, she told me, had suffered a severe stroke and the damage was still being assessed. After several attempts, I was finally able to speak with his wife, Leni. She said—obviously in tears—"He has lost his sight and his memories."*

Julian Schwinger

In the late 1950s, when I was finishing my decade at Harvard, I witnessed an encounter between two titans of modern physics—Wolfgang Pauli who died in 1958 at the age of 58, and Julian Schwinger who died in July 1994 at the age of 76. Pauli was one of the architects of quantum theory. He was also an incredible character—"identical to his caricature," as Robert Oppenheimer once remarked. He had a very large head that bobbed rhythmically as he listened to presentations of physics papers. He was a devastating critic with no tolerance for the mediocre. Of a physicist whose work he didn't think much of, he said "so young, and already so unknown." (My guess is that this was a variant of a Viennese joke—Pauli's natal city—that found its way into the Fledermaus: "So young, and already a prince.") Schwinger—our local titan—one of the founders of modern quantum electrodynamics also had a large head, which he kept characteristically tilted to one side when he listened to physics presentations. But he had a much more kindly disposition. He also had a much more democratic attitude towards other people's papers. He didn't read most of them and rarely cited any. He once told me that it was hard enough doing original research without confusing yourself with other people's ideas.

At the time, Pauli had made one of his frequent visits to the United States from Zurich, Switzerland, where he held his professorship. He was to give a lecture at an American Physical Society meeting in New York and would then come to Cambridge. I was going to the meeting, and Schwinger charged me with inviting Pauli to Harvard for a private lecture session by the Harvard group followed by a reception at the Schwingers. Pauli was not difficult to locate at the meeting, but I was somewhat terrorized by his reputation and barely managed to get out my invitation in what must have been a loud whisper. Pauli had nicknames for everyone. He referred to Schwinger as "His Majesty," and I became known as the "whisperer." (My colleague, the late Feza Gursey, was known as the "Brookhaven Turk." Feza, who was Turkish, had spent a few summers at the Brookhaven National Laboratory.) "The whisperer," Pauli was reported to have said, "has given me a summons from His Majesty to come to Harvard."

The main event at our session with Pauli was, of course, a presentation by Schwinger. But the junior researchers had also been invited to present their wares. I promptly declined this opportunity. I felt that nothing I said would have been remotely interesting to Pauli. I was also aware of his exchange with a Hungarian-born American physicist named Eugene Guth not long before. Guth had made the mistake of interrupting Pauli during a lecture to offer some unwanted bit of erudition. After listening for a moment or two Pauli had said "Guth, whatever you know, I know." There was no way I was going to set myself up for the same treatment. But one or two brave young souls did make brief presentations. Pauli listened patiently and said nothing nasty that I can recall. Then it was Schwinger's turn. Schwinger was famous for his lecture style: He never used notes, formulae came tumbling out of him in cascades, and he did the most amazingly complicated calculations standing at the blackboard with no props. How he managed to keep all of this straight I cannot imagine. As a performance it was overwhelming. Oppenheimer once noted of it "When most people give a lecture they show you how a problem can be done. When Schwinger gives a lecture he shows you that only *he* can do the problem."

Schwinger was then attempting to reformulate the quantum theory, a tricky business to present to Pauli, one of its creators. It was like some artist showing Michelangelo his plans for renovating the ceiling of the Sistine Chapel. Pauli did not interrupt. There was a moment of silence after the lecture, then Pauli said "Wasn't that a little trivial?" I much admired the juxtaposition of "little" and "trivial." I thought I saw a somewhat rictal look flash over Schwinger's usually impassive features—the kind that boxers get when they have received an unexpected blow to the abdomen, followed by that forced smile which is supposed to say "Is that all you've got? It wasn't nothin'." Then Schwinger said "I meant it to be." It was a remarkable answer and it caught Pauli completely off-guard. What could one say? In fact, Pauli found nothing to say and the dialogue stopped there. Both fighters withdrew from the ring.

There is no way of producing genius by education. What one can hope for is an educational system that allows it to thrive. In

Schwinger's case, it was the New York City public schools of the 1930s.*
Schwinger, who was born on February 12, 1918, was the son of im-
migrant Jewish parents who had come to the United States before the
turn of the century. Both families worked in the clothing industry.
Benjamin, Schwinger's father, was a talented designer of women's clothing. Certainly in the years I knew him, Schwinger was a fastidious and
elegant dresser—rather in the Broadway mode. Unlike many immi-
grant families, the Schwingers were always fairly well off. Still, Julian
and his older brother Harold went to Townsend Harris, a public high
school noted for catering to the needs of very gifted students. Two
decades later, the Bronx High School of Science served the same function. In one year it graduated Steven Weinberg and Sheldon Glashow,
who shared the 1979 Nobel Prize in Physics with Abdus Salam.
Weinberg succeeded Schwinger at Harvard when he left for the University of California at Los Angeles in 1972, and Glashow had been a
Schwinger student at Harvard. From 1945, when Schwinger joined
the Harvard faculty, until 1972, when he left for the University of
California at Los Angeles, he had produced more than 60 students, a
great many of whom went on to have very distinguished careers in
physics.

Schwinger's genius for physics and mathematics was recognized
very early. In high school he was already reading advanced papers in
theoretical physics. Schwinger once told me that he had learned a lot
of physics and mathematics by reading the *Encyclopaedia Britannica*—
the old editions that had really serious science articles written by people
like Einstein. In 1934, when he was 16, Schwinger entered the City

---

*While no biography of Schwinger has been written, the interested reader would profit,
as I did, from the following book: Silvan S. Schweber, *QED and the Men Who Made It* (Princeton:
Princeton University Press, 1994). Much of this book requires technical background, but
the biographical details are invaluable.

College of New York—then a tuition-free institution noted for its excellent faculty and its brilliant students. He managed to compile a very erratic record there, failing courses like English. He was spending all his time in the library, reading physics papers and beginning to do serious original work. What would have happened to him is not clear, but his lot was certainly helped by a chance encounter with the Columbia physicist I.I. Rabi.

Rabi, who became one of Schwinger's closest friends, was very fond of telling the story of this encounter. It must have taken place around 1935 after the publication of the celebrated paper by Einstein, Boris Podolsky and Nathan Rosen on the foundations of the quantum theory. In the version Rabi told me, he was trying to understand this paper at the time. His method was to bring a student or junior faculty member to his office and lecture him. In this case he brought in Lloyd Motz, then an instructor at City College and who knew Schwinger very well. At one point they got stuck. Motz then brought into the office what Rabi referred to as a "kid in knee pants." I suspect this was a figure of speech, since I cannot imagine Schwinger wearing knee pants at the age of 17. In any event, Schwinger settled the argument in short order. Rabi was very impressed, so much so that by "using his clout" he got Schwinger into Columbia with a scholarship despite his poor record at City College. In fact, Schwinger managed to make Phi Beta Kappa at Columbia despite failing a chemistry course he thought was too dull to study for. Rabi once told me that Schwinger took a course with the late George Uhlenbeck in some branch of advanced theoretical physics. Uhlenbeck was a visiting professor at Columbia that year. As was his wont, Schwinger never went to class. Uhlenbeck complained to Rabi and Rabi told Schwinger that it was impolite of him not to attend the lectures. I doubt that Schwinger's attendance improved, but Rabi said he was essentially perfect in

Uhlenbeck's final oral examination and even managed to use Uhlenbeck's mathematical notation.

From age 16 on, Schwinger functioned as a professional physicist. When he was 19 he published his first paper, and this became his Ph.D. thesis. It was completed before he received his bachelor's degree. By this time he was the principal theoretical physicist in Rabi's very active atomic physics group. His work habits became legendary. He would appear in the laboratory between four and six in the afternoon, having just awakened. He would then have breakfast while the others were having dinner. Anyone who wanted to work with him had to be prepared to spend the night. In fact, he tried to get Uhlenbeck to give his oral at ten o'clock at night. He was told in no uncertain terms that it would be at ten in the morning.

He kept this nocturnal schedule when he went to Berkeley in 1939 to work with Robert Oppenheimer. They must have made a strange pair. With Oppenheimer, physics was a group activity; he liked to be at the nerve center of this collective energy. With Schwinger it was a solitary activity; he did his own thing in his own way. This was true at the beginning and remained true throughout his career—somewhat sadly at its end when he became more and more isolated from the rest of the physics community.

The work Schwinger did during the period in a variety of fields of theoretical physics was work that any theoretical physicist would certainly have been proud of. It was the sort of thing that eventually found its way into textbooks. The earlier books used to refer to Schwinger. Now the ideas have become so standard that no one remembers who created them. It was not work that would earn Schwinger the 1965 Nobel Prize in Physics that he shared with Richard Feynman and Sin-itiro Tomonaga. That work—the creation of quantum electrodynamics—was done immediately after World War II and used both experimental and theoretical techniques devised during the war, especially in the development of radar.

I would imagine that, because of Rabi's influence, Schwinger chose to move to Cambridge during the war to work on radar as opposed to going to Los Alamos and Oppenheimer. Rabi once told me that he had chosen to work on radar rather than the atomic bomb because he was "serious about winning the war." Without radar he felt we could lose it, while the bomb seemed to him to be something of a long shot. Schwinger developed a general theory of guided electromagnetic waves—waveguides—in cavities of various shapes. In formulating it, he perfected many of the techniques he later used in quantum electrodynamics.

Still he found time to do other physics. After the successful Trinity test in the New Mexico desert on July 16, 1945, Schwinger visited Los Alamos to give a series of lectures on nuclear physics. This was his first encounter with Feynman. The two were almost exactly the same age. This surprised Feynman since Schwinger had already been publishing for nearly a decade. Feynman was still rather unknown to the general physics community, although he was regarded by people who knew him as a genius comparable to Schwinger. Many years later Feynman described his encounter with Schwinger at Los Alamos. He said, in his inimitable way, "The beauty of one of his lectures. He comes in, with his head a little bit to one side. He comes in like a bull into the ring and puts his notebook down and then begins. And the beautiful organized way of putting one idea after another. Everything very clear from beginning to end. You can imagine for a lecturer like me, what a sensation it was to see such a thing. I was supposed to be a good lecturer according to some people, but this was really a masterpiece."*

*S. Schweber, QED and the Men Who Made It, 300.

Both men were so gifted it is hard to compare them. Feynman was, perhaps, Mozartian while Schwinger was perhaps Bachian. In the late 1950s I once watched as Feynman informally lectured a group of his colleagues about a new idea. Schwinger stood off to one side with his head slightly tilted and the hint of a smile on his face. It was a look that was both admiring and skeptical. I wish I had asked him what he was thinking.

Schwinger's masterpiece was his formulation of quantum electrodynamics. This was also Feynman's masterpiece. What then is quantum electrodynamics? First, what is classical electrodynamics? It is the study of electrically charged particles interacting with electromagnetic fields. Since these fields are generated by other electrically charged particles, it is really the study of electrically charged particles interacting with each other. The purest form of such a system is that of electrons interacting with each other and with electromagnetic radiation. The classical theory was created mainly in the 19th century by the Scottish physicist James Clerk Maxwell and then perfected by Einstein. With the advent of quantum theory in the 1920s, new possibilities opened up. Quantum theory allows processes that are forbidden in classical physics. These are processes in which energy and momentum are not conserved—but only briefly. One of the Heisenberg uncertainty relations specifies the amount of time such a process is allowed to take, given that the energy fails to be conserved by some amount. The greater the violation, the shorter the time allowed for it to manifest itself. For example, an electron can spontaneously emit a quantum of electromagnetic radiation—a photon—but then it must reabsorb it rapidly enough so that the Heisenberg energy–time uncertainty relation is obeyed. This is called the "virtual emission" of a photon by an electron. It goes on all the time and is the sort of thing that quantum electrodynamics was created to deal with.

These virtual processes subtly change the properties of the electron. This was well understood in the 1930s when people like Pauli created the original theory. But there were two problems: The experiments done then were not accurate enough to demonstrate these effects unambiguously. Worse, the theory, when applied straightforwardly, gave infinite results—that is to say, total nonsense. That was the real problem. It was left unresolved when that generation of physicists went off to war. After the war, using techniques that had been developed for radar, various physicists finally exhibited these quantum electrodynamic effects in laboratory experiments so precise that they could not be argued away. It was now up to the theorists to make sense of all this. The heros of this story were Schwinger, Feynman and, quite independently, Tomonaga in Japan. It was Freeman Dyson who put it all together and made it accessible to the average working physicist. That Dyson never got a Nobel Prize for this is to me a scandal.

Schwinger and Feynman's approaches were so different that, until Dyson showed the connection, it seemed as if there were two quantum electrodynamics yielding the same results but from totally different starting points. Schwinger began at the beginning. He carefully derived the theory's equations from first principles—isolating the infinite terms. He then solved the equations in a series of successive approximations—a very difficult bit of calculational mathematics. In preparing this essay I once again consulted the collection of papers Schwinger edited on quantum electrodynamics.* In it, he reproduced the third paper in his series of three—"Quantum Electrodynamics III. The Electromagnetic Properties of the Electron-Radiative Corrections to Scattering," which he published in the *Physical Review* in 1949. It gave me the same feeling now as when I first read it some 40 years

*Julian Schwinger, ed., *Quantum Electrodynamics* (New York: Dover Publications, 1958).

ago. It is a massive work; inexorable—like a Bach fugue. But if one took the time—a lot of time—one could follow it. The rules are there. Schwinger also reproduces Feynman's two original papers in which the so-called "Feynman diagrams"—Schwinger used to refer to them as "space-time pictures"—are introduced. These papers also gave me the same feeling I had when I first read them. They are full of wit, and delightful mathematical tricks abound. But I recall that I felt I would never really understand them. What justified the tricks? It was Dyson's two papers—also reproduced in the collection—that showed that Feynman's diagrams and Schwinger's calculus were two ways of representing exactly the same mathematics.

What really counted in these papers were the numerical results. Schwinger got there first. He computed a tiny change in the electron's magnetic interaction—its so-called "magnetic moment." [I always thought that "magnetic moment" would be a great name for a perfume.] This number has been refined again and again over the years, and it has been measured with ever-increasing precision. It is now the most precisely known quantity in physics. A recent experimental value is:

$$\frac{1}{2}(g-2) = 1159652182.5(4.0) \times 10^{-12}$$

while the theoretical value is:

$$\frac{1}{2}(g-2) = 1159652140(28) \times 10^{-12}$$

The numbers in parentheses represent error estimates. The agreement is remarkable.

When I entered Harvard in 1947 as a freshman, I was totally unaware of any of this and cared less. Over the next three years I made a series of Brownian motions through various majors, finally ending up in mathematics in my senior year. My guru was the mathematician George Mackey. Mackey had originally been trained as a physicist and had become interested in trying to formalize—as mathematicians like to do—the mathematics of quantum theory. It seemed like an interesting program, and one that I might like to work on except that I knew next to nothing about quantum theory. Mackey suggested that I take Schwinger's course, which purported to be an introduction to the subject on the first-year-graduate level. By this time I had at least heard of Schwinger from my physics-major colleagues, but had never set eyes on the man.

My first impression of "Julian," as he was known to all his students (who wouldn't have dared call him that to his face), was that he was "larval." He had the pallor one associates with those asparagus the French grow underground—pale, for lack of sunlight. He was certainly overweight. A wraithlike Indian graduate student who was sitting next to me, and who would have been blown away in a zephyr, commented (*w*'s became *v*'s) "Our Schvinger is *very* fat." He wasn't *that* fat—just plump. The class was nominally scheduled for sometime after 11:00 *a.m.*, thank God. At the time, Schwinger drove a small blue Cadillac. When it was spotted we all took our seats. In he would walk, as Feynman said, "like a bull into the ring." He had no notes and as a rule made no mention of anything that had been said in the previous lecture. He just began at the beginning and raced on to the end. One had to write like mad just to keep up with what was being put on the blackboard.

I had, it turned out, taken this course at a fortunate moment. Schwinger was just beginning his work on reformulating quantum theory. Therefore, he began the course by reexamining all the experimental reasons why there was a quantum theory in the first place. This was an exercise in phenomenology involving almost no formal mathematics. It was a lot harder to do than solving equations, as I found out years later when I began teaching quantum theory. Schwinger was a master of this as well. I still have the notes. I have never seen a better introduction to the subject, but in the middle of the year Schwinger began doing his own formalism and lost a great many of us. I, for one, went down to MIT to audit Viki Weisskopf's course, which was much more down to earth. However, many of the ideas that Schwinger was developing have since become part of the standard literature of our field.

After taking my master's degree in mathematics, the math department made it clear that I was spending too much time taking physics courses. I would have to make up my mind to be either a doctoral candidate in physics or mathematics. I chose physics and have never regretted it. I wanted to get started on my Ph.D. thesis. It would be a mistake, I thought, to try to work with Schwinger. He already had many students, and they knew a lot more physics than I did. Typically, they got a few minutes every week or so to discuss their thesis problems. Many said that during the two or three years they worked on their theses they might have seen Schwinger for all of 20 minutes. I figured that I might need at least 20 minutes a day! I chose, therefore, to work with a young postdoctoral researcher named Abraham Klein, who has been at the University of Pennsylvania for many years. Klein had been a Schwinger student, so that put me in the next generation— the 'grandsons' of Schwinger.

It was only after I took my degree, and became a postdoctoral at Harvard myself, that I got to know Schwinger. There was such a small group of theoretical physicists at Harvard and MIT that we all knew each other. In fact, once a week we would meet for lunch after Schwinger's lecture in one of the Cambridge restaurants where we could all fit around one table. Julian—as I now called him to his face—used to try out his latest ideas on people such as Weisskopf. I saved, and still have, one of the paper napkins on which he had written some equations. I remember the atmosphere. It was very exciting and, as it turned out, rather delusional. The reason was that the kind of physics needed in this new world of elementary particles was not really Julian's forte. The problems were not very well posed. What was needed was a kind of speculative intuition—not Julian's kind of formal mathematical power. A new generation of physicists—people like Murray Gell-Mann, T.D. Lee and C.N. Yang—became our intellectual leaders, later to be succeeded by people like Steve Weinberg and Sheldon Glashow. Julian did not seem to fit into this very well. The physics on my napkin, while perhaps not wrong, was not really relevant to what was going on.

Just after I left Cambridge in 1957, Julian seems to have undergone a life change. He had gotten married a decade earlier to a very attractive brunette by the name of Clarice Carrol. He had subsequently given up smoking. This was a good thing for more than one reason. It was reported that while he was doing his work on quantum electrodynamics in his office one night, he dropped a lit cigarette into a paper-laden wastebasket. A serious conflagration was avoided because a charwoman noticed the smoke and doused it. But in 1958 Pauli died at age 58—very young by our standards. This appears to have affected Schwinger a great deal. Like Schwinger, Pauli was also very overweight and underexercised. It seems as if Pauli's early death motivated Julian to become a model of physical fitness. I wish I had been around Cambridge to see the transformation. I recall being an object of some ridicule in my day because I rode bicycles, rowed and played tennis. After Julian went to UCLA in 1972, I was told that he played tennis with a pro almost every day. Certainly, the few times I saw him during those years he was tanned and looked very fit indeed.

I am not sure how happy Schwinger was during these years. Rabi once told me that Schwinger had bitterly complained to him that the younger generation no longer took him seriously. Colleagues at UCLA told me they never saw him. I am sure this isolation was self-imposed. Julian was doing his own thing in his own way. But one cannot have it both ways. Einstein also did his own thing in his own way—rejecting quantum theory to the end. But he didn't care what the younger generation thought; that was *their* problem. Given a different mind-set, Julian might—based on a lifetime of research—have written one of those multivolume texts that sets the whole field's tone for a generation. As it was, he did publish a few books. But, as usual, they were his own thing in his own way. It was not clear what the advantages were of doing things his way, and few people bothered to sort out what he had in mind.

During those years I sent Julian and Clarice the odd Christmas card and occasionally got an answer—always from Clarice. Once she even sent me a reprint of a paper that Julian had found while cleaning a closet, which he thought I might like to have. Not being a California person, I never did have the chance to visit them. The last news I had about Julian, before his unexpected death in July 1994, was that he had become interested in "cold fusion." Someone sent me a reprint of a lecture Julian had given in Japan on the subject. I read it, but I kept wondering why Julian had ever gotten into this unfortunate subject.

In writing this appreciation I am well aware of what I cannot convey, especially to the younger people who were not there. It is all very well to say that someone was a genius or a fantastic lecturer. But how can one re-create the feelings that we had about people like Schwinger. Our generation, and the generations that followed, have certainly produced extraordinary physicists and remarkable teachers. I have known many of them. But I still feel that in the days when people like Schwinger and Pauli were at the height of their powers, giants walked the Earth.

LEPTONS

A t this point I would like to beg the reader's indulgence. The first of
these lighter—"leptonic"—fictions—"Bubble and Squeak," has already appeared in a much earlier collection of mine. I am including it here for two
reasons: First, I wanted the opportunity to rewrite it some. I like this version
better than the previous ones. But more importantly, it will give me an ex-
cuse to describe a brief interaction that I had with The New Yorker
magazine's fiction department. This was the old New Yorker—The New
Yorker of William Shawn.

At the time I began writing for the magazine in the early 1960s, there
was an established tradition of writers who wrote both "fact"—in the
magazine's jargon—and fiction. The names Dorothy Parker, E.B. White,

Edmund Wilson, John Hersey, among others, come to mind. My generation included people like John Updike—who was a "Talk of the Town" reporter—Calvin Trillin, Ved Mehta, Renata Adler, and many others who worked in both genres. The writing I had done before I began writing professionally in the 1960s, consisted of poetry and short stories. I did this for my own pleasure, just as many of my colleagues played the piano for theirs. I don't recall showing anyone this writing, and I certainly never tried to get it published.

However, a few years after I had begun to write for the magazine on a regular basis, I resumed writing short stories—actually "casuals," New Yorker lingo for a short, often humorous bit of writing. But now I had a sounding board—William Shawn, the magazine's editor. I suppose I must have submitted two of these casuals a year for him to look at. They were all promptly rejected. About once a year I had lunch with Mr. Shawn at his table at the Algonquin Hotel. After two or three years of unsuccessful fiction writing had gone by, Mr. Shawn took it upon himself to deliver, at our annual lunch, a little homily on fiction writing. Much of it was unsurprising. He noted that in fiction it was both legitimate—and indeed often necessary—to use stylistic devices that, at least as he conceived New Yorker fact writing, were impermissible. For example, a major no-no in New Yorker fact writing was "indirection"—the introduction of people or other items from left field. When Mr. Shawn's predecessor, Harold Ross, encountered unexpected individuals in fact pieces, he would query on proofs "Who he?" In fiction, on the other hand, such a surprise might be quite appropriate.

However, Mr. Shawn's parting advice on fiction writing was quite unexpected. He said that for fiction to be really successful, the writer must hold back no part of him- or herself. He added that if one could do that, it would be a source of "release"—I think that was the word he used—for the writer. During the weeks that followed this lunch I thought a great deal about what he had said. I realized that even my fact articles gave almost nothing away about myself. They were very impersonal. There was a reason for this. I was writing mainly about scientists who were both considerably older and vastly more distinguished than I was. I felt that my role was to convey their ideas and not my own. But I was also writing many unsigned "Talk of the Town" pieces in which I was much less austere. Perhaps that was why Mr. Shawn thought that if I could bring the same skills to writing fiction it might work out.

Some months later I got the germ of an idea for a short story. I had been reading a fair amount about René Thom's "catastrophe theory"—the Theory for Everything that was then in vogue. It has subsequently been followed by fractals, chaos theory, and now, complexity. What struck me about Thom's theory—and the others that followed it—is how radically they had to modify the universe in order to describe it. This is true of all abstract science, but the pretensions are generally fewer. I had just had an unhappy love affair, and I began to wonder which of Thom's mathematical "catastrophes" described it. The notion struck me as incredibly funny, and out of it came "Bubble and Squeak."

But there was something else. When I was a sophomore at Harvard my first great teacher in physics and its philosophy, Philipp Frank, introduced me to the writings of Wittgenstein. We worked through the Tractatus together. I was then, and am still, quite unclear as to what exactly Wittgenstein was trying to say, but I was much struck by his style. As readers of the book will remember, it consists of numbered paragraphs, like a mathematical logic text. This gives it an aura of misplaced rigor—like applying catastrophe theory to a love affair. Hence, when I wrote "Bubble and Squeak," I numbered the paragraphs. I did not include the quotation from Wittgenstein, which I have added to this version.

This piece of writing was so different from anything I had done before that I did not know what to make of it myself. Still, I submitted it to Mr. Shawn. Then a curious thing happened. There was no response. Usually, the rejections had come within a few days; now a couple of weeks had gone by. My curiosity got the better of me, so I called Mary Painter, Mr. Shawn's secretary. Mary was the soul of discretion, but when she sensed that a writer was becoming desperate she would sometimes be willing to give a few hints. In this case the message was mixed. Mr. Shawn, she said, liked my story. Indeed, she liked my story—but the fiction department didn't. Mr. Shawn, she explained, was trying to change their minds. I had no idea the fiction department had such autonomy, but apparently it did. On rare occasions, I was told, Mr. Shawn would overrule their decision and publish something anyway—but not often. A few days later, Mr. Shawn called and confirmed the situation. He also told me that he had not been able to convince them, and would not be able to publish it. One of the things that bothered them was the use of the numbered paragraphs. (I guess no one there had read Wittgenstein.) Eventually, it was published in the now-lamented Mountain Gazette, a marvelous small magazine bankrolled for a while by my Aspen friend George Stranahan.

This near miss persuaded me that I might actually produce an acceptable piece of fiction for The New Yorker. But what about? Here, in Pasteur's phrase, "chance favors the prepared mind." During my many years as an academic I attended very few faculty meetings. Indeed, I had as little as possible to do with academic politics, which I detested, but it was difficult to escape entirely. What I saw of it was either extremely depressing, or hilarious, depending on whose ox was being gored. I witnessed things that went beyond fiction. Finally, I reacted by creating my own imaginary university. The thing that worried me, I used to say, was that I might get tenure there. I didn't, but three pieces of writing emerged, two of which were published in The New Yorker—the only fiction I ever succeeded in publishing in the magazine. Then I more or less stopped writing fiction, although I still have occasional malicious thoughts about my universities—real and imaginary. Now I am a professor emeritus at both.

Bubble and Squeak

1. *Die Welt ist alles, was der Fall ist.*
1. *The world is everything that is the case.*

L<small>UDWIG</small> W<small>ITTGENSTEIN</small>

A mathematician cannot enter on subjects that seem so far removed from his preoccupations without some bad conscience. Many of my assertions depend on pure speculation and may be treated as daydreams, and I accept this qualification—is not a daydream the virtual catastrophe in which knowledge is initiated? At a time when so many scholars in the world are calculating, is it not desirable that some, who can, dream?

R<small>ENÉ</small> T<small>HOM</small>

1.   I knew a mathematician who had a recurrent dream. He dreamt that he was a partial derivative. The number of people who

have dreamt *about* the differential calculus could be quite substantial, especially if one includes engineers and economists. However, only a professional mathematician could dream that he had *become* the differential calculus.

2.   The British have always had a way with ghosts. As far as one can tell, this has to do with atmospheric density. Ordinary matter has a density of one to ten grams per cubic centimeter. A laboratory plasma has a density of about 10^{-8} grams per cubic centimeter. The density at the center of the sun is about 10^2 grams per cubic centimeter, while neutron stars have a density of about 10^{15} grams per cubic centimeter. I estimate the average ghost density at about 10^{-4} grams per cubic centimeter. This is a tentative estimate. Ghosts cannot be made in a vacuum. That would violate the conservation of matter. If a ghost is created on the surface of a neutron star, it can never escape. The gravitational attraction is too strong. On the other hand, the probability of ghost formation in the Sahara desert—if my calculations are right—is extremely small: about one ghost per three hundred years per cubic centimeter. The atmosphere over Great Britain—taking into account both Scotland and Ireland—is quite dense. I estimate that one ghost per three minutes materializes somewhere in the British Isles. This accounts, no doubt, for the frequently reported sightings by reliable, and often disinterested, observers.

My friends Richard and Sally Longwood have seen a ghost. They live in London. Sally believes that it is the ghost of her great aunt who was French. "She spoke to me in French," Sally has told me. She, herself, does not speak French but she certainly recognizes it when she hears it spoken. "Pity," says Sally, "I would so like to know what she was trying to tell me." That is why I am here in the Longwoods' dining room. I do speak French and have agreed to interpret. The dinner table has been cleared and the lights extinguished. There is now a candle burning in the center of the table. The three of us are seated around the table with our hands resting lightly on it. "Is anybody there?" Sally asks, politely, several times. There is no response. Then Richard tries—no answer. Then Sally tries again—with no success. Finally, she says "I don't think my aunt will materialize tonight. Perhaps she is shy with people who are not family."

3.   Sometimes I am invited to dinner if I am willing to take "pot luck." The French call it *"pot au feu"*—a *potpourri*—a "rotten pot" of things boiled over a fire. The British call it "bubble and squeak." "Do drop over for dinner—that is, if you don't mind a bit of bubble and squeak." Usually I don't mind.

4.   A mathematical theory of catastrophes must be rigorous. A "catastrophe manifold" must be defined and "control parameters" must be introduced. "Energy functions" may be plotted with a ruler and straightedge. Books and articles can be written, and conferences can be addressed. Awards can be handed out. Tenure can be achieved. According to Thom, if there are at most four control parameters and two behavior variables, only seven basic catastrophes are possible. These include the "butterfly," the "swallow tail" as well as the more complicated "elliptical," "parabolic" and "hyperbolic" umbilics. The simplest catastrophes can be described in terms of cusps. Cusps come in handy. They can be plotted in three dimensions. The catastrophe manifold takes unexpected twists and turns and, if one is not careful, one can fall over the edge. Anxiety and frustration can suddenly turn into anger. Hate and love form a continuum, and one can hardly tell the players without a score card. The brain is an interconnected network of billions of neurons, to say nothing of axons and dendrons. If one is

not careful, it can develop cusps. At times, going for a long walk is a good way to deal with them. At other times, it can only make things worse.

5.   There are two single women in this hotel. They do not seem to know each other. Each is accompanied by a child. In the absence of *a priori* arguments to the contrary, it is reasonable to assume that each woman is the mother of the respective child that accompanies her. There is also genetic evidence that can be brought to bear in a pinch. The blonde woman has a blond child, and the black-haired woman has a brown-haired child. Why is this convincing evidence? Well, we know that natural hair coloring is not an acquired characteristic. You can't get it from a bottle. No matter how often you dye your hair, your child's hair will not be purple. Hair coloring is inherited. It does not grow on trees. But—genetically speaking—it takes two to tangle. There is more here than meets the eye.

I have gathered some additional evidence concerning the blonde woman. She leaves the hotel each morning at 9:30 with the blond boy. (I forgot to say that the blonde woman is accompanied by a blond *boy* while the black-haired woman is accompanied by a brown-haired girl. This detail would not have changed any of the arguments much. The sexes appear to be distributed more or less at random.) The blonde woman always wears slacks and, frequently, a heavy-knitted white sweater. We are in the mountains. She has very good posture. In fact, she is almost rigid. She never smiles in the morning. Each night she has dinner with the boy at the same table in the dining room. She drinks exactly one-half a bottle of red wine, and the boy drinks one glass. We are in France. Then her face becomes flushed and she smiles a great deal. Each time she smiles, the boy smiles too. I do not know if the boy smiles when he is alone. (Speculation without experimentation can rapidly produce cusps.) They always converse with their neighbors at the next table. The conversation appears to be very animated. But the occupants of the next table change every few days, while the conversation remains the same. After dinner, the blonde woman and the blond boy say goodnight very solemnly and disappear. This has gone on night after night. One day I happened to be standing at the reception desk when she was mailing a letter. It was addressed to someone in England. I smiled at her, but she did not smile back. I wish I could tell you more about her. One cannot always get blood from a writer's imagination. Sometimes one has to settle for bubble and squeak. We will now turn our attention to the other woman. Perhaps we will have better luck.

She appears every other day at the swimming pool of the hotel with her daughter. This deserves further study. Why every *other* day?

6. Frustration dreams must be common in every culture. But how are they expressed? Mine often involve public transport—taxis, buses, trains and the like—which pass me as if I were invisible. What do Sherpas and Zulus dream about? Or Eskimos. Snow and ice? What would an Eskimo frustration dream be?

7. There *have* been technological breakthroughs in long division. There is no denying that. It used to be, with the old mechanical desk calculators, that if you divided a number by zero the machine would grind on forever—trying to reach infinity. Sometimes smoke came out of the gears. Now, if I use an electronic calculator—the one I use to compute the ghost probabilities—and divide, say, one by zero (1/0), it will give me the largest number it can think of: $9.999999999 \times 10^{99}$. Then it starts to blink. Sometimes I divide zero by zero just to see what will happen. It then produces 1 and starts to

blink. It does not seem too certain of the answer otherwise why would it blink like that? This too deserves further study.

8.   The black-haired woman with the brown-haired child has just arrived at the pool. She is wearing a bikini. She has what people often call a lovely figure. The use of the phrase "lovely figure" in this context has always puzzled me. As far as I am concerned, a lovely figure is something like 88 or 111. It would, I think, be more appropriate in *this* context to say that she has a magnificent body. She owns at least three bikinis. The argument for this is easy to follow. I have seen her three times at the pool, and each time she has been wearing a different bikini. God knows the number of bikinis that she actually possesses. The bikini that she is wearing today has the color of fine Burgundy wine. It has been manufactured, one would gather, out of a shimmering, water-resistant substance that closely resembles the skin of mermaids. Her exposed skin, of which there is a great deal, has a marvelous amber tone, as if it is being illuminated from within. Each time I have seen her, she has settled on the opposite side of the pool, directly across from me. I have no way of estimating the probability that this is accidental. She spreads out her towel and lies down on it, on her stomach, with her feet pointing towards the pool. From where I am, I estimate that the angle between her calves and the back of her thighs is about eight degrees. This is just an order-of-magnitude estimate. It may have to be revised as we go along. The child, who must be about eight, wears a bikini bottom but no top. The places where her breasts will be have a gentle convex shape—like that of a miniature Japanese sumo wrestler. As quickly as possible she gets into the water.

9.   Insofar as there is such a thing as theoretical biology it should concern itself with options. After all, sex is optional. By this I mean

that the fact that there are two—and only two—sexes appears to be entirely arbitrary. There *are* alternatives. Take the female greenfly. She can bear live, fatherless female offspring identical to herself. Further-more—and get this—the daughter is born with an embryo for *her* daughter already inside her womb—daughter and granddaughter are born simultaneously. Genetically they are identical twins to their mother and grandmother, respectively. Not much is known, at least to me, about how greenflies feel about this, but I have heard that male biolo-gists, after having seen greenflies give birth a few times, are known to leave their laboratories and head for the nearest bar for a stiff drink. I have a mathematician friend who spent some time working on a theory of n-sexes, where n could be *any* positive integer. He was trying to prove that there was some distinct advantage in having $n = 2$, which he referred to as the "classical case." He never found an argument that really satisfied him. At about this time, his wife went off with a stock-broker. I must find out what he is working on now.

10.   After getting into the water, the child swims back and forth five or six times tracing an imaginary line that connects me to her mother. If her mother and I were flowers, and the child a bee, one would refer to this activity as "pollination." Finally, the child stops at the side of the pool where I am lying. She has done this each time she and I have been at the pool together. Then she begins to tell herself a story. She does this in a low voice, but loud enough so that I can hear. She does not look directly at me while she is telling herself the story, but every once in a while she glances in my direction to make sure that I am listening. I am.

Her stories always involve playing with a group of imaginary friends. I feel that these friends are imaginary because their names keep changing from day to day and from story to story. Sometimes these friends invite her to play when she is having dinner or about to go to bed. There is then an adult—always male—who tells her that she cannot play now, but can see her friends tomorrow. The adult male is never given a name in these stories. I do not feel that the child wants me to speak to her just yet. Rather, I am a witness. My main role is to indicate in one way or another that I am listening. After a while, the child's mother turns over and calls "Nathalie," and the child swims away.

11. Creativity, it has always seemed to me, violates the conservation of matter. Where have the ideas *been?* If the average ghost density is only 10^{-4} grams per cubic centimeter, what is the matter density of Nathalie's imaginary friends and father, to say nothing of a lot of other ideas I could mention. Some scientists I know think that the quantum theory will have to be modified to take "ideas" into account. But isn't the quantum theory itself an idea?

12. Nathalie and I have begun speaking to each other. She has commented to me about the temperature of the water. Since I never go into the water, I am unable to confirm or deny her assertions about it. While I am quite willing to concede that the temperature of the water is a numerical parameter that we should be able to measure with thermometers and the like, what instruments should we use to measure our *feelings* about the water? Even if Nathalie and I were to agree on the numerical value of the temperature, we might still disagree about our feelings as to whether or not it is desirable to swim in it. I would, of course, respect her point of view even if it differed from my own. That is one of the burdens we take on as we grow to middle age.

Whatever Nathalie and I have to say about the water has now been said. On the other hand, for some time I have been puzzled by a riddle—or, more exactly, by the solution to a riddle. I know the riddle, but not its solution—a common human dilemma. I used to know both the riddle and its solution, but now they have become disconnected in my memory. As I recall, both the riddle and its solution are absurd. It is the connection that is interesting—if only I could locate it. The riddle is "What has a hooker, a looker and two sticky wickys?" This, I believe, uniquely specifies something. But what? Nathalie has theories—even conjectures and educated guesses. This sort of thing is just the ticket for scientific research, where progress is rarely made in leaps or bounds. But with riddles, what is required is a leap, or perhaps a bound. Nathalie and I have wracked our brains—whatever that means. Perhaps "raked" our brains would be better. We have hit a stone wall. In desperation I suggest that perhaps Nathalie's mother could be brought in, on a *per diem* basis, as a consultant. Nathalie swims over to her mother and explains the problem. Nathalie's mother turns over on her back, looks at me and smiles. Is this a situation that can be described by the quantum theory? Perhaps, but words are better.

13.   The probability of two people falling in love is difficult to compute—even with the best computers. In probability space, we may represent each independent probability as a circle whose area is proportional to the probability in question. If the two events whose probabilities we represent by circles are independent, then all we have to do is add up the areas of the two circles to find the resultant probability; if the two events are *not* independent then we must find the area of the arena where the two circles overlap. I could do this readily on my pocket calculator if I had the foggiest notion of the areas of any of the circles involved—to say nothing of where they overlap. It is, it would seem, considerably more difficult to estimate the probability of two people falling in love than it is to find the probability of a pair of dice adding up to seven. This is a domain where theory is not going to get us very far. We must proceed empirically.

14.   A fly has developed in the ointment. After smiling at me, Nathalie's mother has gotten up and gone into the hotel. She has not returned and neither has Nathalie. The sun is setting. Since we are in the mountains, it has gotten quite cold. I will send them a note inviting one, or both, of them to share something—an activity perhaps. It is all very good and well to have decided on this, but as a practical matter I have very few names to go by. In fact the only name I have is that of Nathalie. The issue of where to send the note, however, is resolved by presenting the problem in outline form to the desk clerk. Since there are only two unaccompanied women staying at the hotel—one blonde and the other black-haired—we are able to place the correct woman (from my point of view) in a well-defined hotel room, which she shares with Nathalie. The desk clerk has also told me Nathalie's mother's last name. It is "La Farge." First names are of no practical importance to him, professionally speaking, so he has no knowledge of Madame La Farge's first name. This raises the dilemma of how I should address the note. To begin it with "Cher Madame La Farge" would sound as if I am either looking for a job or trying to collect a debt. The only solution that occurs to me is to not begin it with anything. I begin the note by simply identifying myself as unambiguously as I can, considering the rather limited number of coordinates we have in common. I then propose taking tea in the near future, sign my name and give my room number. I then leave the note in the appropriate mailbox.

15.   A second fly has now developed in the ointment, which it shares with the previous fly. I have left my note in the appropriate mailbox after dinner. I conjecture that the answer—if there is one— will be delivered to my mailbox the following day. Unfortunately, I will not be in the hotel the following day. I will be in the mountains. Guides have been engaged. Crampons and ice axes have been sharp-

ened. The ropes that will attach specific guides to specific climbers have been tested for tensile strength. The weather will be excellent. In short, there is no turning back.

16.   Conditions are abominable. Great gaping holes have appeared in the glaciers. Ordinarily dry rocks are covered with sheets of ice. Avalanches appear to be imminent. "Things fall apart: the centre cannot hold. . . . " No one knows quite why this has happened. It certainly was not foreseen. The guides appear to be concerned. We are instructed to concentrate on the task at hand. My mind, however, wanders. Among the subjects under active consideration are the following: I have estimated empirically that the angle between the back of Madame La Farge's thighs and her calves, when her legs are fully extended, is about eight degrees. Is this a universal angle—an angle that is the property of women in general, or is it a special property of Madame La Farge? In dealing with this matter, strict empiricism has its limitations. Although the total number of women who now exist or who have existed is, one would think, finite, no complete tabulation is available. If there were such a thing as theoretical biology it would begin with a few general principles; universal angles—if there are any—would come out in the wash. As it is, we are reduced to rumor and speculation—traditions handed down from father to son. The problem is further complicated by custom and usage. In some societies these matters are deeply felt and rarely discussed, while in others, the reverse appears to be true. The light is not visible at the end of the tunnel, and furthermore, the particular guide to whom I am attached has begun to pull smartly on the rope. I shall return to these matters at the earliest opportunity.

17.   Madame La Farge's first name is "Veronique"—at least the note that I found in my mailbox was signed "Veronique La Farge." It is

unlikely that she would have made up a first name just for purposes of this note. There is little point, I would say, in presenting a line-by-line translation into English of this note, which, by the way, was written in a flowing, bold handwriting on hotel stationery. The translation might well lack something that the original possessed. Let us instead deal with the idea, or ideas, contained *in* the note. In fact, since the note itself consisted of a single sentence our task is relatively light. Veronique La Farge and Nathalie will meet me in the hotel *after* dinner to discuss the tea. The fat, in brief, is on the fire.

18. The reading matter in the salon of the hotel seems to have been selected so as to obey the "law of the excluded middle"—*tertium quid non datur.* On the one hand, there was an eight-month-old copy of *Jours de France,* which featured a full-scale, one page, black-and-white photograph of Brigitte Bardot water-skiing in San Tropez. Miss Bardot is shown on the skis waving at something, or someone, with her right hand. In her other hand she is holding a pair of ropes which are attached to a large motorboat. The boat is being driven by a thin, but muscular, young man with a great deal of hair. Both people are smiling, but not at each other. Their thoughts have not been recorded. On the other hand, there was a thick industrial review that contained several thoughtful articles dealing with the production of metal tubes in various countries in Western Europe. Both the tubes themselves, and their methods of production in the various countries, are carefully compared. Photographs of the tubes and the factories in which they are produced are displayed. There is also a brief, but clearly written, article on the annual gross national product of Liechtenstein.

I had begun to speculate on the minimum size that a political entity must become before it *had* a gross national product and had tentatively reached the conclusion that Liechtenstein might actually be somewhat below this theoretical minimum when Veronique La Farge and her daughter entered the salon. Both women were wearing very long skirts. I cannot, at this time, recall the color of Nathalie La Farge's skirt, but that of her mother was a deep oceanic blue. She had on a richly brocaded white blouse that closed, on top, with a large pin. Her black hair rested symmetrically on either shoulder. I also noticed that she was wearing a small wristwatch with a filigreed gold band. She also had on a large engagement ring with a blue center and a diamond circumference. When we shook hands, I noticed that the ring was being worn on the wrong finger.

The two women sat down on a small, green divan across from my chair. They smoothed down the fronts of their respective skirts with nearly identical sweeping motions. Veronique La Farge spoke first. "I have received your note," she said. This was clearly a statement on which we could agree.

Many Asian societies have long realized that by preparing and consuming endless cups of tea, the consideration of vexing, perturbing and possibly even irrelevant questions can be postponed or even deflected completely. In the situation that now confronted me I decided that the reverse might apply. Hence, I asked Veronique La Farge if she had made any progress in locating the solution to my riddle. "No," she replied, she had not, but she was still thinking about it. Perhaps the root of her difficulty lay in my defective French translation of "sticky wicky." After all, "walkey-talkey" is translated into French as "talky-walky," which proves that the unexpected lurks around every corner when it comes to translation.

I then asked what Veronique La Farge and her daughter did on those alternate days when they did not appear at the pool.

"We go for walks in the mountains," she replied.

"Do you have friends here in the valley?" I asked.

"No," she said.

I then proposed that the three of us go for a walk in the mountains on her next alternate day, which I knew was the next day since this afternoon, when I returned to the hotel, I had caught sight of her at the pool. She agreed, and a time and a place were arranged. After a few minutes of additional conversation she got up, as did Nathalie. We shook hands and they left the salon. Only then did I remember that we had forgotten to discuss the tea.

19.   Goals must be set and priorities must be established. This appears to be true throughout the animal kingdom. It is a natural bodily function and certainly nothing to be ashamed of. After a certain amount of reflection, I decided that a suitable goal for our hike would be to have lunch. This goal had at least two advantages: it was attainable, and once having been attained, it would be generally recognized that it *had* been attained. Most of the other goals that I had considered I found to be defective in one, or both, of these respects. After reaching this decision I telephoned Veronique La Farge and told her that I would provide the lunch for the two of us if she would provide the lunch for Nathalie. While I am familiar with the component parts of the digestive system of an eight-year-old child, and even have some understanding of how these parts are interrelated, I have found as yet no method of putting this information to use to decide what the child would enjoy digesting.

This dilemma is not entirely unanticipated. There are two schools of thought: There are those who say that the existence of poetry follows logically from the existence of dictionaries, and there are those who are not quite so certain. To be sure, both dictionaries and books of poetry exist and can be purchased in bookstores or borrowed from libraries. What cannot be purchased or borrowed, it would seem, is the connection between them. In making this connection, bubble and squeak has, no doubt, a role to play, but it is difficult to believe that it can be the whole story. A good deal of the research remains to be done.

20.   I have prepared an itemized list of the contents of my rucksack:

1 tin of pâté de foie gras

2 pork sausages

1 cold chicken

2 small quiche Lorraine in a cardboard box

1 loaf of white bread

2 varieties of cheese

1 pocketknife with three blades

1 corkscrew

3 chocolate bars

1 small cake

3 oranges

1 bottle of red wine

1 bottle of mineral water

1 flashlight with an extra battery

3 cloth napkins

1 pair of extra shoelaces

1 small first-aid kit

6 paper cups

1 rain jacket

1 wool hat

1 magnetic compass

2 maps

The rucksack including these items weighs approximately thirty-one pounds. In order to accommodate these items I have had to remove from my rucksack a plastic climbing helmet, a pair of crampons, a coil of rope and five pitons. Hopefully, none of these will be needed.

21. The lunch has been a great success. Nathalie has described, in considerable detail, her school. It is located near a large park in Paris. On Tuesdays and Thursdays after school she attends a ballet class with her friend Justine. Justine is afraid of the dark, and why not?

After lunch, Nathalie takes a doll from her small rucksack and goes under a tree to take a nap. "She always takes a nap in the afternoon," Veronique La Farge informs me. Veronique La Farge has told me about her divorce. If Veronique La Farge's marriage can be characterized by four control parameters and two behavior variables then, according to Thom's theorem, her divorce can have been only one of the seven basic catastrophes. I am having some difficulty in deciding which one. Perhaps it was one of the umbilics. It sometimes takes decades before a theory can be usefully applied. This can lead to some impatience, and an increase in cusps will be noticed in the general population. Only in retrospect do we come to understand that what looked like a cusp at the time was really a molehill. Veronique La Farge has also noticed the blonde woman with the rigid posture and the blond boy. She assures me that this posture cannot be acquired by ice-skating, since her former sister-in-law was an ice-skater with ordinary posture. From time to time, Veronique La Farge tells me, she and Nathalie receive a letter from her former husband, who has moved to Canada with his current wife. Nathalie saves these letters in a cardboard box.

In the difficult spots on the path back down to the hotel, Nathalie holds my left hand. On the *very* difficult spots, Veronique La Farge holds my right hand. The whole, in this instance, is considerably greater than the sum of the parts.

22.   The light of the afternoon sun coming through the Venetian blinds of my hotel-room window casts a parallelogrammatic shadow on Veronique La Farge's naked back. She is sleeping on her stomach, and my hand is gently resting on the back of her left knee at the place where her calf joins her thigh. As she wakes up I say to her "Veronique, you are very beautiful." Sometimes things that we predict do not come to pass, and at other times things come to pass that we have not predicted.

23.   Even bubble and squeak has its constraints. Insofar as it takes place at all, it does so in space and time, and, if some recent theories of the universe are to be believed, these are commodities that we may be running out of. Furthermore, there are the conservation laws—energy, momentum, angular momentum, baryon and lepton number, to name a bare minimum. While these are not as deeply understood as one might hope, they must, in any case, be obeyed on a daily basis. Summer turns into fall. Decisions must be made and later regretted. There is nothing to be done about this. It runs throughout the animal kingdom. Veronique La Farge and Nathalie have left the hotel. I have put them on the train to Paris. Schools must be attended and ballet lessons must be resumed. Perhaps we will all meet again. Meanwhile I have received a note from Sally Longwood in London. Her great aunt—the one who spoke only French—has once again materialized in the Longwood dining room. She is desperately trying to communicate something. But what? Tomorrow I leave for London.

The
Philosophy
Circle

Since early this morning I have been taking propositions from various treatises by Wittgenstein, typing them onto neat white sheets of paper, and then gluing the squares in a column on a large sheet of cardboard. I have also glued an "antisquare" adjacent to each square on which I have typed the same proposition with all the verbs negated. Here is a sample:

The world is everything that is the case.
The world is not everything that is not the case.

If there were no connection between the act of expectation and reality, you could expect a nonsense.

If there were a connection between the act of expectation and reality, you could not expect a nonsense.

Language must speak for itself.
Language must not speak for itself.

And so on.

Tonight the Philosophy Circle is meeting at my house. Last week we met at Alice Dodd's apartment. The subject was "Alternatives to the Law of the Excluded Middle." Alice Dodd's middle has been excluded, at least to me, ever since she started sleeping with Herbert Feist, the department chairman. Before she took up with Feist she was a logical positivist. Feist is a Kantian idealist—*Ding-an-sichlich*, and all that. I am a logical positivist, and during the meeting I insisted that the law of the excluded middle was either true or it was not true. Tonight's subject is "Does the Philosophy of Wittgenstein Have a Content?" That is why I have been gluing my little squares. My idea is that if Wittgenstein's statements and their negations both seem equally true or untrue, then they have no content and we can go on to something else—anything else. I have also prepared the cheese and wine.

The doorbell rings. It is old Professor Lash. His main claim to fame is that he studied for a term in Cambridge when Wittgenstein was still alive. Once, when old Lash was drunk, he confided to me that while he was there Wittgenstein refused to talk with him. For some reason Wittgenstein couldn't stand him. I tried to console him by saying that if Wittgenstein were still alive he would be appearing regularly on the *Tonight Show*. "Heeeeere's, Ludwig!" I said.

Old Lash was scandalized. "Wittgenstein hated interviews," he muttered.

"But that was before television," I pointed out.

I helped him home, but before he got out of my car he recited a limerick he said he had heard from G.E. Moore. I have always liked Moore's comment after reading Wittgenstein's thesis: "It is my personal opinion that Mr. Wittgenstein's thesis is a work of genius; but, be that as it may, it is certainly well up to the standard required for the Cambridge degree of doctor of philosophy." The limerick, though, was not much good.

Pith! That's what it takes to get ahead in this world. There is no such thing as too much pith. Take the last maxim in Wittgenstein's *Tractatus*: "Whereof one cannot speak, thereof one must be silent." Have you ever read such pith? Tonight that will be the password. Pith. No more disruptive remarks like the one I made at the last meeting, when I said that our meetings reminded me of the partial inverse of Lord Acton's maxim.

"Whatever do you mean?" old Lash was good enough to inquire.

"It's that absolute lack of power corrupts absolutely—if you take my point," I explained.

Feist, who was presiding, looked a little liverish, but didn't say anything. Alice Dodd's eyes flashed the way they do when she gets angry. There will be none of that tonight. Let bygones be bygones. Water over the dam. Tonight I am going to be as pithy as the grave.

"White or red, John?" I say after I have taken Lash's coat. Once he has decided on a color for his evening's wine, he likes to stick with it.

"Red, if you don't mind," Lash replies.

The doorbell rings again. It is Alice Dodd. Her cheeks are a healthy rose, and her long blond hair is neatly tied in a ribbon. She is wearing a tartan wool skirt in deep reds and greens. I think it is new.

She looks around my living room. "George, you seem to be doing interesting things with your furniture," she comments.

"Yes," I reply. "I have sold most of it."

Before she can say anything, old Lash, who has already been working on the wine asks "How is the book coming?" She and Feist are "bringing out"—or perhaps "bringing up"—a monograph on Kant's last major essay "The Conflict of the Faculties." When Feist mentioned this to me I told him that I had never read it because I had always assumed that it was a discussion of academic politics. "While I have always admired Kant's sense of humor," I added, "I have never been entirely certain that it would lend itself to a pastiche on academic life. Did he also write light verse? You can tell so much about a philosopher from his light verse. Hegel had a wonderful touch." Feist did his thesis on Hegel.

The doorbell rings again. It is Albert Backen. Poor Albert is not tenured. Alice Dodd and I, in the days when she was still a logical positivist once made up a plainsong that began with the verse:

Poor Albert hath no tenure.
No ten-y-ure hath he,
No ten nor ure no ten nor ure hath he.

I am fond of Albert, but I worry about him. He wears his lack of tenure on his sleeve like a black band of mourning. "Come in, Albert," I say. "I am pleased you could come. We need young minds like yours." (Actually, Albert has never missed a meeting.) He looks up at me brightly. Perhaps I have heard something from the promotion committee. Old Lash interrupts. "Once when I was visiting Wittgenstein in his rooms in Trinity, he confided to me that. . . . "

"John," I say before he can get any further, "I read recently that the only films that Wittgenstein would see were westerns. He felt a special affinity with Tom Mix. Did he ever mention that to you?"

The doorbell rings again. This must be Feist. Warren Drake, the only other member of the department to come to our meetings, has gone to Nevada for a divorce. I open the door. Bless my soul—it is Feist, and what a nice new sports jacket he is wearing. "A present from your wife?" I inquire loudly, so that Alice Dodd will hear. "We were just discussing Wittgenstein's obsession with Tom Mix. I think that John was about to point out that this may have had something to do with Wittgenstein's sexual preference." (I have always liked the phrase "sexual preference." It conjures up in my mind an image of one of those Korean greengrocers where the dear legumes are all tarted up in colors so bright that they look as if they have just come fresh from the embalmer. "What is your preference in lettuce?" asks the kindly grocery person.)

"Wittgenstein *had* no sexual preference," Lash remarks.

Before anyone has a chance to inquire further, Feist says "Sorry I'm late. I've just come from the dean's office. Burning the midnight oil, you know."

"Well put, Herbert," I remark, noting out of the corner of my eye that poor Albert has turned white at the mention of the dean. "Is there anyone here that you haven't met?"

Feist arches his eyebrows slightly but does not say anything. He greets Alice Dodd a little formally and poor Albert hardly at all.

Feist takes his chair across the room from Alice. I am about to bring out my sheet of cardboard squares, but before I have a chance to do so Albert begins to talk. "There is a passage in Wittgenstein's *Philosophical Remarks*," he says, smiling winningly at Feist, "which has been giving me a good deal of trouble. Professor Feist, perhaps you could elucidate it for me?"

285

Oh Albert, you poor bastard, I think. Feist couldn't elucidate the meaning of a telephone bill.

"I would be pleased to," Feist says condescendingly.

Albert reads, after pointing out that the passage is on page 110 of the little blue paperback edition we all have in front of us, "'I haven't got a stomach-ache' may be compared to the proposition 'These apples cost nothing.'" He hesitates, expecting that Feist will have something useful to say. Nothing is forthcoming.

Albert continues reading, "'The point is that they don't cost any money, not that they don't cost any snow or any trouble.'" He hesitates again. In fact, he has come to a dead stop and is peering intently at Feist. I can see that Feist has been taken completely off guard and is stalling for time, in the hope, perhaps, that old Lash will say something and rescue him. But Lash has now drunk nearly a full bottle and is not to be counted on.

"I should have thought . . ., " Feist begins.

"Quite so," I interject.

"I should have thought," he goes on, "that within the general Hegelian warp and woof. . . . "

Suddenly old Lash sits up and says "I believe the expression is *Warf und Woof*," after which he subsides seraphically back into his corner.

Alice tries to come to his rescue. "I think there is a misprint in the text. It should read, 'They don't *cause* any snow or any trouble.'"

"I am not sure that will wash," I say, "unless you are prepared to change the second proposition so that it reads, 'These apples cause nothing.'" By now I can see that I have tossed pith to the winds.

Albert, who appears to be entirely oblivious to what is going on, and who, poor sod, really *wants* to understand Wittgenstein, adds "I have gone through the *Philosophical Remarks* with some care, and I have underlined all of Wittgenstein's references to apples. Here is one on page 132."

We all turn to page 132 and Albert reads "'If I have 11 apples and want to share them among some people in such a way that each is given 3 apples, how many people can there be?'" Once again he looks at Feist and draws a blank. He reads on: "'The calculation supplies me with the answer 3.'"

I can see that Lash is collecting his thoughts and is about to say something. "I think that John wants to make a point," I interrupt.

"Wittgenstein was very partial to fresh vegetables," Lash explains, "But only in season."

There is a dead silence.

Albert, who somehow has the idea that it is his responsibility to say something, begins again. "When I was at Harvard . . . ."

Oh Christ, I think. The doomed son of a bitch really has a death wish. Doesn't he realize that Feist hates Harvard? The graduate school turned him down, and he had to do his graduate work in the Midwest.

Feist's right fist is clenched white. As dense as Albert is, he does appear to notice that something has gone terribly wrong.

"When I was in graduate school," he continues hesitantly, "Quine once told us . . . ."

Before he buries himself completely, old Lash comes unexpectedly to the rescue. "Young man," he says, gesturing with a half-filled wine glass, "it's very much like comparing apples and oranges."

God knows what old Lash has in mind, but Albert interprets this sibylline comment as a request for more passages dealing with fruit. He seems to have made a horticultural reading of Wittgenstein's entire opus. "Professor Lash," he says respectfully, "here is something about oranges in Wittgenstein's *Philosophical Remarks* that you might find interesting. On page 276, toward the bottom of the page. . . . " We all turn our blue books to page 276, "Wittgenstein writes 'Admittedly it's true that we can say of an orange that it's almost yellow, and so that it is "closer to yellow than to red" and analogously for an almost red orange. But it doesn't follow from this that there must also be a midpoint between red and yellow.'"

"Exactly the sort of thing I had in mind," Lash comments.

I sneak a glance at Alice Dodd. In the days when she was still a logical positivist we would come back after a session like this to my house, collapse on what was then my sofa, and laugh until we were in tears. "This too shall pass," I would say, echoing a comment I once made about a group of sophomores who were taking my survey course in modern philosophers. Now she is sitting upright in her chair. Her face is a mask. God knows what she is thinking. Feist is staring intently at the ceiling.

Albert goes on. "Here is something I came across last night in the *Philosophical Grammar.*" He takes a fat red paperback volume out of his briefcase and opens it to a page he has marked with a thin sliver of paper. "'What does the process or state of wanting an apple consist in?'" he begins.

What does the state of wanting Alice Dodd consist in? I think.

"'Perhaps I experience hunger or thirst or both, and meanwhile imagine an apple, or remember that I enjoyed one yesterday. . . . Perhaps I go and look in a cupboard where apples are kept. Perhaps all these states and activities are combined among themselves and with others.'"

Feist has shifted his stare from the ceiling to his carefully polished left shoe. I am sure it has been polished by his wife.

Albert continues inexorably, "On page 140 of the *Philosophical*

Remarks, just above the diagram, Wittgenstein asks 'Can I know there are as many apples as pears on this plate, without knowing how many? And what is meant by not knowing how many? And how can I find out how many? Surely by counting. . . . '"

We all stare uncomprehendingly at the diagram. It resembles the skeletal structure of an especially rigid fish, and seems to have wandered onto the page by accident.

"Do you happen to have any apples?" Albert asks.

"Yes," I reply. "I think there are a few in the kitchen, but they are a bit past their prime." I was planning to give them to the department secretary.

"I need them to illustrate the next example, which is rather abstract," Albert explains.

I go into the kitchen and am rummaging about in a heap of uncertain fruit, attempting to locate the remaining apples, when the kitchen door opens and then closes. It is Alice Dodd.

"Look at this curious hole," I say, pointing to a perforation in one of the apples I have managed to unearth. "Do you think that whatever made it was going from the inside out or the outside in?"

"Why did you sell your sofa?" Alice Dodd asks.

"I found it distracting," I reply, "The pattern on the slipcover was much too busy."

"I rather liked it, " Alice Dodd remarks.

My, my, I think. What have we here? Perhaps Alice Dodd is finding that Kantian idealism is wearing a bit thin. But before I can explore the matter in more detail the kitchen door opens again. It is Feist. When he catches sight of the two of us his face turns the color of those extraordinary *asperges* that the French manage to grow entirely underground. When they dig them up they look like little shrouds.

"Herbert," I say affably, "Professor Dodd and I were just ducking for apples. Perhaps you would like to join us."

"I . . . ," Feist begins.

"No need to apologize, Herbert," I say. "Take a handful."

We return to the living room, apples in hand. Old Lash is now dozing off. That is the last we will hear from him. I will have to ask Albert to take him home. The four of us divide the apples into little piles, and Albert begins to read: "'If I say: If there are four apples on the table, then there are 2 + 2 on it, that only means that the four apples already contain the possibility of being grouped into two and two, and I needn't wait for them actually to be grouped by concept' . . . "

At this point we each take four apples and group them into two and two. Feist is staring intently at his apples, although from time to time he sneaks an anxious look at Alice Dodd.

Albert continues. "'This *possibility* refers to the sense, not the truth of a proposition. 2 + 2 = 4 may mean "whenever I have four objects, there is the possibility of grouping them in 2 and 2."'"

"Albert," I ask, "do you think the logic would work equally well with six apples? Are *four* apples absolutely essential to the argument?"

"I haven't tried it with six," Albert acknowledges.

I glance at Feist. He looks as if he is about to explode. The thought crosses my mind that perhaps he is going crazy. It would do wonders for the department.

He suddenly gets up to leave and, with a desperate look in the direction of Alice Dodd, announces that he has an early-morning appointment with the dean.

"Before you go, Herbert," Alice Dodd says, "we should settle on our next meeting. I propose that we have it at my place again and that the subject should be 'Do We Exist?'"

"Of course, of course," Feist says, and he leaves without saying goodbye.

Well, they are all gone now. Alice Dodd and I helped Albert pack old Lash into Albert's car. Alice gave me an affectionate wink when she got into hers. Now, alone among the shards of decaying apples, I can take stock of the evening. On the minus side, I was never able to display my little tableau of Wittgenstein and anti-Wittgenstein propositions. No matter—I will use it in my course next spring. On the plus side, there is clearly a marked softening in Alice Dodd's Kantian idealism. Perhaps I can repurchase my sofa from the Japanese mathematician I sold it to. The slipcover can always be changed to a somewhat less-aggressive pattern.

As I am collecting the last of the apples, I notice that Albert has forgotten to take his annotated copy of Wittgenstein's *Philosophical Remarks*. In fact, he has left it open to page 64. Sure enough, near the top of the page there is another reference to apples. Wittgenstein writes "If I wanted to eat an apple and someone punched me in the stomach, taking away my appetite, then it was this punch that I originally wanted."

What an odd thought. What could Wittgenstein possibly have had in mind? I must ask Albert in the morning.

━━━━━━━━━━

Until I published this story, I had not realized that The New Yorker's noted fact-checking department checked fiction as well as fact pieces for factual accuracy. The first question I was asked was whether or not I had made up all these quotations from Wittgenstein. I was able to show them that each and every one was just as the master wrote it—page numbers and all.

Portrait of Bleibermacher

Each summer, Maurice Bleibermacher prepares his annual lecture. "The twick, Geowge," Bleibermacher, with whom I went to graduate school—before he acquired his present accent—once confided in me, "is to skim the cweam." Freely translated, this means that each spring Bleibermacher peruses the list of Guggenheim winners and then makes up a lecture from the titles of the successful proposals. It is a brilliant idea. It is to lecturing what the late Bernie Cornfeld was to mutual funds. Cornfeld had a fund that had only shares of other funds. It had the leveraging power of a hydrogen bomb—in both directions. Last year's lecture, which I had the misfortune to hear, was called "Freud, Quantum Mechanics and the Collective Unconscious—Where?" The

"Where?" was Bleibermacher's masterstroke. The main point of the lecture seemed to be Bleibermacher's attempt to show that the photon was conscious. Some of the physicists in the audience were persuaded that Bleibermacher did not know the difference between a photon and a watermelon, but before they could lay a glove on him he managed to change the subject to a discussion of the influence of the Upanishads on Franciscus de Marchia's notion of the *virtus derelicta*. If a loaf of Wonder Bread could lecture, it would sound like Bleibermacher.

Even when we were graduate students Bleibermacher had acquired a great fondness for terms like "Poisson Brackets" or "Pirus Japonica," to say nothing of "cayita-vrksas"—none of which he understood. They rolled off his tongue like Three-in-One oil. About this time, Bleibermacher married an unfortunate, mousy girl named Janice who supported them by working as a secretary in the physics department. She and Bleibermacher had gone to high school together somewhere in Ohio, near where Bleibermacher had been born. Later, after the divorce, he gave his birthplace, when asked, as Salzburg. He had, by this time, acquired a curious foreign accent that no one could quite place. He had come to the conclusion that all successful British intellectuals, no matter what their mother tongue, pronounced "robot" and "realism"—"wobot" and "wealism." They also stuttered and squinted—mannerisms that Bleibermacher adopted as well. After spending time in his company, I always found that my lips began to pucker involuntarily.

After graduate school, Bleibermacher grew a goatee and added fifteen pounds. This only enhanced his sex appeal. Women told me that he gave off a mysterious sexual energy—like a quasar. From time to time, I would encounter him at meetings of the Modern Language Association or the American Philosophical Society with one or another incandescent young blonde whom Bleibermacher would introduce as his secretary. They all looked as if they had just been washed ashore on the Windandsea Beach at La Jolla. They hung limpetlike on Bleibermacher's every word. So I was not overly surprised when I received the following letter:

Dear George:

As you probably know, I shall be coming to you for convocation. [Posters with Bleibermacher's picture were already hanging all over the campus.] Perhaps you can give me some advice. [Now what?] I cannot quite decide as to whether to present my ideas on cosmology or entropy. [Translation: whether to give this year's lecture or last year's. He was probably worried that a number of people, besides myself, had already heard last year's.] I am, of course [The use of "of course" in what followed was another of Bleibermacher's better touches.], struck by the congruences between Snorri Sturluson's account of the origin of the universe in the *Younger Edda* and the *Popol Vuh* of the Quiché Maya. Have you ever noticed the striking similarities between the Unkempt and Black Sorcerers and Nifflheim and Muspelheim? Is this yet another striking example of continental drift? Was Iceland ever part of Guatemala or was it vice versa? Was Einstein aware of the parallels? Can we regard his cosmological constant as still another realization of the giant Ymir?

On the other hand, I have recently come to the conclusion that the second law of thermodynamics has been overemphasized in the California school system. I made a survey of the junior class at Laguna Beach High School and discovered that about 60 percent of the boys had recurrent dreams about the impending heat death of the universe. I was able to reassure them by quoting a theorem of Poincaré. You recall the one that states that if you wait ten billion years a scrambled egg will reassemble itself into a hen. It's just a matter of time.

On a more personal note [Here it comes.]. President and Mrs. Praeger have been good enough to invite me to stay in their home during convocation. It appears that this is a tradition. I find, therefore, that I will have to book a room somewhere, not too far from the campus, for my secretary, Seleina Marlow, who has kindly agreed to accompany me. Can you suggest a motel?

I value your views.

Maurice

I sent Bleibermacher a postcard with the telephone number of the local Holiday Inn.

I managed to lie low for the next few weeks while the final preparations for Bleibermacher's visit were being made. I heard that Praeger had come around to the department to look for me. This came as something of a surprise, since, as far as I could tell, Praeger never seemed to know which department I was teaching in. He kept trying to speak to me in Portuguese. Finally, Alice Dodd learned the reason for the sudden interest and reported back. Somehow Praeger had discovered that Bleibermacher and I had been to graduate school together, and he needed my help in clearing up a dilemma. The library had wanted to organize a little display of Bleibermacher's books, but it had not been able to find any. Praeger thought that I might have some first editions. "Bleibermacher has never written a book," I told Alice Dodd. "He once told me that the messenger was the medium."

Finally the great day arrived. A note had come down to the chairpeople from Praeger to the effect that he wanted each department represented at the airport by two members. In a handwritten aside to Feist, he added that because of my "close" relationship with Bleibermacher, I was to be one of the representatives of the philosophy department. I told Feist that as much as I would like to go with him, I had a prior appointment with some graduate students who wanted to learn more about Mill's view on induction. "They want to know," I explained, "how many white swans we have to see before we can say with reasonable certainty there is no such thing as a polka-dot swan. I feel that they are entitled to an answer." As soon as I saw Feist leave in his car for the airport, I went home and poured myself a stiff drink.

Later that evening the phone rang. My heart sank. I thought it might be Bleibermacher or possibly Praeger or Feist. I was quite confident it wasn't a student since I keep my phone number unlisted to avoid those callers. I answered the phone. Fortunately, it was Alice Dodd. Although she and Feist are barely on speaking terms after their affair, she had been in her office when Feist caught her. He had insisted that for the good of the department (everyone else had evaporated) she come to the airport with him. She was phoning to give a report. Bleibermacher, she said, got off the plane with an incredible blonde in tow—Seleina Marlow. Praeger was thoroughly bewildered since, as he put it, he "had not been informed that there was a Mrs. Bleibermacher." The guest room the Praegers had prepared, it appears, had only a single bed. The provenance of Seleina Marlow was explained to Praeger. Then there was a general exodus back to the campus in various cars.

Somehow, Feist managed to capture Bleibermacher, Seleina Marlow and Alice Dodd in his car, so Alice Dodd was able to observe what ensued. "Feist was determined to score points with Bleibermacher," Alice Dodd reported. He probably wanted an invitation to Bleibermacher's place in California, but Bleibermacher, with all his faults, was hardly about to spend his hard-earned foundation money on the likes of Feist. "He," Alice Dodd went on, referring to Feist, "started an argument with Bleibermacher as to whether neutrinos ex-

297

perience free will." It was unlikely that either Feist or Bleibermacher had the foggiest idea what a neutrino was. "Bleibermacher," Alice Dodd went on, "was all over Feist like a shroud. It was like watching the slaughter of the innocents. He managed to bring in both the Demotic Mathematical Papyri and Huang Ting's *T'ien-wer-ta-ch'eng kuan-kuei chi yao* in a single sentence. Feist looked as if he was about to have a hemorrhage. He was reduced to muttering 'I should have thought' over and over. Just before they came to blows—which would have been interesting—the Seleina Marlow person piped up and asked to be taken to the Holiday Inn. When we got there, she and Bleibermacher got out. He announced that he would take a taxi to the Praegers since he had some urgent dictation for Ms. Marlow. Feist dropped me off without even saying goodbye." So much for Feist's chance to visit Big Sur.

I have been giving careful consideration to the following mini-max problem, as they say in the theory of games. How can I minimize the number of my actual contacts with Bleibermacher, while at the same time maximizing the number of times Praeger sees me in Bleibermacher's general vicinity? Feist has been moving heaven and earth to see that Praeger does not give me a raise next year. If Praeger gets the impression that Bleibermacher and I move in intersecting orbits, it might help matters. It is crucial to this strategy that Bleibermacher and I don't start on each other within earshot of Praeger. A simple exchange of banalities will suffice. The first major test of this plan will be at the Praegers' reception for Bleibermacher this evening. To calm my nerves I read, as I often do in circumstances like this, the medical entries in the *Encyclopaedia Britannica*. I have just finished a fine paragraph on gangrene—"a synonym in medicine for mortification or a local death in the animal body due to interruption of the circulation by various causes."—when I realize that I am already late for the Praegers'.

When I arrive, I discover that to get to the drinks it is necessary to go through a receiving line that consists not only of Praeger and Bleibermacher but also Lenora Praeger. (Seleina Marlow is nowhere to be seen. Perhaps she is back at the Holiday Inn typing.) Lenora Praeger looks like one of those figures one makes by pressing fresh dough with a cookie cutter and then putting in raisins for the eyes. In her case, kumquats would have been better. She is the organist in the college chapel. She has persuaded Praeger that it would be a lovely idea if the faculty would attend services to set an example for the undergraduates. At every faculty meeting Praeger makes an appeal for chapel attendance. It is heeded by a few ancient professors such as old Lash (the local bars are closed on Sunday), and junior faculty members like Albert Backen who think this might improve their chances for tenure. Naturally, the students all go to New York for the weekend, which is what the rest of us would do if we could afford it.

There are a few people in front of me in line, so I have a chance to study Bleibermacher. He is wearing a very expensive three-piece suit with a vest that floats on top of his paunch like a Tibetan prayer flag. I think it is the same one he wore for his last appearance on Letterman. He is elegantly tanned and is wearing tinted aviator's glasses. The goatee is still there, but something seems slightly out of place. For a few minutes I can't quite place it. Then it strikes me. Bleibermacher has had a hair transplant. What used to be occasional grayish tufts on the top of his skull have been replaced by youthful-looking sprouts of black hair, which seem slightly out of place. The whole effect is like one of those geological inversions in which the younger Triassic rocks somehow land on top of the Jurassics leaving for the geologists the puzzle of explaining how they got there. In Bleibermacher's case, the little sprouts of black hair look like the remains of a somewhat erratic harvest of a wheatfield.

When I get to Bleibermacher and Praeger, Praeger gives me a suspicious look. Before he can say anything I say to Bleibermacher, in as amiable tone of voice as I can manage, "How is Trixie?" Praeger looks startled. Trixie is the name of the dolphin that Bleibermacher has borrowed from the San Diego Sea World for some sort of experiment. It would be much more intellectually rewarding if Bleibermacher were lent *to* Sea World. "The wowk is getting vewy intense Geowge, vewy intense," Bleibermacher answers in his inimitable accent. I had read somewhere that the unfortunate animal was showing all the signs of an acute nervous breakdown and that Sea World was thinking of bringing suit. Despite my strategy, the temptation to ask Bleibermacher about this was all but overwhelming. But before I could say anything, Bleibermacher went on. "I was telling Walter [a reference to Praeger] what a dedicated graduate student you were—always writing. What are you writing now? We miss your little essays. The style was so light—like a fairy's kiss."

That does it. The bastard has already done me in. Feist must have told him about the abortive monograph on the sexual excesses of Rudolf Carnap that I have been writing for the last seven years. Feist brings it up whenever I ask for a raise. No need for any further strategy. The cause is lost, so I say to Bleibermacher "I like the way you are doing your hair Maurice. Do you use a combine or a simple thresher?" Praeger, as dimwitted as he is, seems to have become aware that the conversation is rapidly getting out of hand and he says to me, rather urgently, "Try some of Lenora's punch, George. She spent the whole afternoon preparing it." "I may have to pass it up, Walter," I say. "As you know, I am allergic to fruit." Praeger wheels Bleibermacher off, and as soon as I can, I make my way home. I drink a third of a bottle of Cointreau and read myself to sleep with the *Encyclopaedia Britannica* entry "Skin, Diseases of."

The next noon—Saturday, convocation day—I get up with a splitting headache and decide to go to the campus coffee shop for some black coffee. There, needless to say, is Bleibermacher. Sitting next to him, and listening with rapt attention, is Seleina Marlow. From where I am sitting it looks as though she is crocheting something. It resembles a net. Perhaps it is for Trixie. Also listening in rapt silence is a group of undergraduates. I find a table as far away from them as possible, but one of the students comes over to report that Bleibermacher is explaining how the Sanskrit treatise—the Khundakhadyaka of the Brahmagupta—proposes to calculate the manda of Saturn. "A very important concept," I comment, not having the remotest notion of what, if anything, the manda of Saturn is, "and well within your intellectual capacity to understand," I add. The student, looking a little puzzled, returns to Bleibermacher's explanations, and I go home and read the *Encyclopaedia Britannica's* entry on "Joints" to pass the time before the convocation, which has been scheduled for the early evening.

I had hoped there might be a tornado that would blow away the tent, preferably with Bleibermacher inside it, where the convocation address is to be given. But no such luck. It is a windless evening, and the tent is filled to capacity. A portable organ has been set up and Lenora Praeger, dressed in a kind of white sail with pink bows, plays some approximation of Brahms' "Academic Overture" to get things started. In fact, she might have gone on indefinitely if Praeger had not succeeded in prying her away from the organ. A few awards are given out and some retirements are announced. Old Professor Lash is retiring at the end of the year. Two of the junior faculty have to help him to the podium. He has spent the afternoon in the Faculty Club bar with a bottle of port. As he weaves his way uncertainly back to his seat, I can hear him muttering, "But they promised me a watch." Then there is Bleibermacher's speech preceded by an interminable introduction by Praeger. It was like water-skiing in treacle.

Bleibermacher had decided to give this year's lecture—the one on entropy. He begins by taking a brightly colored sheet of paper and tearing it into shreds, which he throws at the audience. "This is an increase in entropy," he announces. "You have just witnessed an actual increase in entropy." There is a stunned silence in the tent, and then applause and cheering. Praeger is in such a state of rapture that, for a moment, I think he is about to cry. Next, Bleibermacher announces that a new codex of Leonardo da Vinci has been discovered in Elizabeth, New Jersey, showing a clear drawing of an outboard motor. This is such a non sequitur that I wonder, for a moment, if Bleibermacher has actually lost his mind. But I observe him rapidly shuffle his lecture cards like a blackjack dealer in Las Vegas. He must have gotten one of them from last year's lecture mixed in. No one else seems to have noticed, and he makes a seamless transition back to entropy—the nominal subject of his discourse.

By this time I have settled into a kind of meditative stupor from which I am suddenly aroused by Bleibermacher's mention of Maxwell's demon. I used to think that this was a drink favored by the Scots until a physics colleague of mine explained that Maxwell had invented it to show that the second law of thermodynamics was wrong. The demon stood in the middle of the container of gas, selecting all the fast molecules for his own purposes. His portion of the gas went from cold to hot without anyone being the wiser.

Bleibermacher stood in the middle of the lecture platform selecting molecules from thin air and shouting "Après moi le déluge!" which must have been part of last year's lecture on cosmology. The audience gave him a standing ovation. Then Praeger called for questions. One undergraduate noted that in a recent issue of *Omni* magazine there was a photograph of the back of one of Leonardo's sketches for the Mona Lisa that revealed, in the master's hand, a perfectly executed drawing of a bowling alley. Bleibermacher replied that he was well aware of the discovery but hadn't mentioned it because of lack of time.

There are a few more questions. Then Praeger announces that there will be one more. Much to everyone's surprise, old Lash struggles to his feet. He grabs the back of the chair in front of him and shouts "I should have gotten a Bulova! That's what they usually get." There is a stunned silence. Bleibermacher looks genuinely bewildered. But then he pulls a small metal triangle out of one vest pocket and a little rod out of another. He must have these with him, just in case. He taps the triangle with the rod, holding it next to the microphone. The tone reverberates throughout the tent. He breathes deeply and says "Om" several times and sits down. The lecture is over and we all file out.

Six months have passed. I did not get my raise. Praeger has stopped speaking to me altogether. I have heard nothing from Bleibermacher, although I did see him on the *Today* show. Trixie is back at Sea World and is apparently well on the road to recovery. Bleibermacher is now trying to teach a chimpanzee named Simba to use Windows 95. I thought I was well rid of Bleibermacher, but just this morning I received a short note from him.

Dear George:
Seeing you meant a great deal to me. In fact, I want you to be the first to know that President Praeger has offered me the Herschel J. Lang Chair in Cognitive Studies, which I have accepted

Incidentally, could you look around near the campus for a house that I might buy? Don't worry about the price.

I value your views.
Maurice

The
Faculty
Meeting

Alice Dodd was attempting to resuscitate her philosophy class. Through the open door I could see students in various stages of rigor mortis. Irving Nafken, for example, was wearing that rictal grin, which has often given me the urge to place his head gently but firmly between Volume VIII (Poy-Ry) and Volume IX (S-Soldo) of the Oxford English Dictionary. He once asked in a class of mine whether it was true that Schopenhauer was a "fruit." "Schopenhauer's life was not entirely happy," I replied. "His father drowned himself in a canal, and Schopenhauer was sued by a seamstress who claimed he had beaten her with a stick. In court, he said that they were having a discussion on the reality of will and that he was merely trying to make a metaphysical point. He lost the suit." Another student, Edwina Gwan, was studying her eyebrows in a hand mirror. She once came to my office dressed in a miniskirt and a see-through blouse, ostensibly to discuss original sin. I told her that most sins seemed already to have been invented by someone else—or perhaps by a committee—and that there should be a Nobel Prize for the inventor of a really original sin. Soon after, she began sleeping with our department chairman, Herbert Feist.

By shifting my position slightly, I could see that Alice Dodd had written something on the blackboard. I recognized it as one of the paradoxes invented by the logician Willard V. Quine:

"Yields a falsehood when appended to its own quotation"
yields a falsehood when appended to its own quotation.

The thing that has always puzzled me is how Quine ever thought up a sentence like that. I once wrote a poem that began "Willard V. Quine is smarter than I'm. . . . " I never finished it.

Alice Dodd spotted me, but before she could say anything the bell rang. This was just as well, because Nafken had his hand up. Perhaps he was about to ask if *Quine* was a fruit. He once told me that he intended to become a brain surgeon and that he was working his way through school by repairing color television sets. Perhaps he saw a connection—all those brightly colored wires.

I entered the rapidly emptying classroom.

"George, what in God's name are you doing here?" Alice Dodd asked. "Don't you have office hours now?"

"Look," I said, "don't bother me with trivia. This is a matter of life and death. Bleibermacher's got loose again."

Bleibermacher has made an entire career out of comparing things—any things—none of which he understands. For example,

306

his latest paper, which was largely written by a graduate student, was called "Bell's Theorem and the Shu-shu chiu-chang." The latter, I learned from Bleibermacher, was a treatise on algebra and number theory written by the 13th-century Chinese mathematician Ch'in Chiu-shao. I told Bleibermacher that it was the "and" in his title that troubled me. "Despite," I felt would work somewhat better. "Geowge," he said, in his inimitable accent, "yaw twabble is that you have no sense of humaw."

To give him his due, Bleibermacher has turned comparison into an artform. He once showed me a file drawer full of little cards emblazoned with titles of articles that he or his graduate students intended to write. Things like "Reichean Orgones and Solar Electricity" and "The Slime Mold and Its Use in Indian Art." Each one was a rich fount for grant proposals.

Indeed, Bleibermacher has managed to parlay all of this into a remarkable career. Until he came to our place he had never held an academic job. He didn't have to. Foundations tripped over each other in their rush to support him. Guggenheims, Sloans, Revsons, MacArthurs came his way one after the other. What they thought they were supporting God only knows. I suppose each one reasoned that the last place that had given him a grant must have known what it was doing. It all reminded me of a man I know who found himself stateless after the Second World War. His wife stitched together a document that looked like a passport. He said the real problem was getting the first visa. Some country—I think it was Liechtenstein—gave him one, and after that he got visas from everywhere.

In recent years Bleibermacher had been operating out of Big Sur, but he had a few contretemps there. A "wild child" whom he claimed had been raised from infancy by a herd of elk near Aspen turned out to be a teen-age runaway from Santa Barbara. Her family reclaimed her after she appeared with Bleibermacher on the David Letterman show, walking on all fours and making odd barking sounds. Not long afterward, the dolphin that Bleibermacher was trying to teach to do the crossword in the London *Times* had a nervous breakdown and had to be returned to the San Diego Sea World.

In fact, Bleibermacher might have fallen on hard times had it not been for the providential appearance of Herschel J. Lang. Lang is a self-taught biologist who has made a great fortune in clones. He first came to the attention of the general public when he accidentally produced a species of bacteria that would eat only Chinese food. The original strain was allergic to MSG, however, and it was in the course of trying to improve the breed that Lang happened on the discovery that made his fortune. He was feeding a batch of his bacteria some of Admiral Tsai's Historic Bean Curd when they began producing a substance that was, for all intents and purposes, indistinguishable from the blend of Dacron and rayon used in three-piece suits that sell, new, for under $89.95—alterations extra. The rest, of course, is history. Lang's company, Splice-Tech (its motto is "They grow so you can sew"), is now among the Fortune 500, and he recently entered the world of *haute couture* with a modified strain of *E. coli* that produces mauve crêpe de Chine.

One might have thought that with all this worldly success Herschel Lang would have been a happy man, but he suffered from a monumental inferiority complex. He had no academic degrees. "I can buy them and I can sell them," Lang often said, referring to biologists with Ph.D.s, "but they still have their god-damned lambskins." In truth, Lang never graduated from high school. His father had been a *tummler* at places like the Concord and Grossinger's, and he trained his son to follow in his footsteps. "To *tummel*," he often told young

Herschel, "you need timing—and timing they don't know from at Harvard." So Lang had been denied a formal education, something he had never been able to reconcile himself to. Enter Bleibermacher.

The two of them had met on Letterman, where Lang was exhibiting a strain of bacteria so specialized that it would eat only shredded pork with Peking sauce prepared by the chef at the Lop Sum restaurant in Cleveland. Bleibermacher, on the other hand, was exhibiting a trained gerbil, which he claimed could take cube roots with the aid of a specially designed calculator. Lang saw in Bleibermacher a man heavy with degrees and light in cash; in Lang, Bleibermacher saw the reverse. After a certain amount of diplomatic small talk, Bleibermacher got down to cases. He would locate a not-overly-scrupulous university president—a quantity in generous supply—who would allow Lang to endow a chair in, say, cognitive studies, on the condition that Bleibermacher be its first occupant. Once Bleibermacher was firmly in his chair, he, in turn, would begin a campaign to have Lang awarded an honorary degree. Lang would have his lambskin and Bleibermacher would be set for life.

As it happened, Bleibermacher and Lang did not have to look very far. The president of our place, Walter Praeger, was a devoted talk-show viewer and had become a Bleibermacher fan. He somehow found out that Bleibermacher and I had been to graduate school together, and after that, took every opportunity to pump me for information. I tried to explain to Praeger that if Bleibermacher ever had a really clear idea, it would kill him like an electric shock. But Praeger would have none of it, and last spring he invited Bleibermacher to be our commencement speaker. By the end of the weekend, Praeger and Bleibermacher were on a first-name basis, and not long thereafter Lang's limousine could be seen pulling up to the campus regularly. It came as no surprise, therefore, when in the fall Praeger announced the creation of the Herschel J. Lang Chair in Cognitive Studies, whose first occupant would be, needless to say, Maurice Bleibermacher. The only difficulty was that neither Bleibermacher nor Lang had bothered to inform Praeger until it was too late that part of the deal was an honorary degree for Lang. That was what I was about to tell Alice Dodd.

"He's got to be kidding," she said. "As depraved as Praeger is, he hasn't stooped to selling honorary degrees."

"It looks like he has," I said. "Perhaps, even as we speak, the public-relations office is beginning a sales campaign with ads in *Scientific American* and *The Wall Street Journal:* 'Small but prestigious university has honorary degrees for sale. Terms to be arranged.' Those people are capable of anything."

"But George," Alice Dodd replied, "you know as well as I do that an honorary degree has to be approved by a vote of the faculty."

She was right. I had completely forgotten. Honorary-degree approval, along with mandatory early retirement, was written into our contract after we were unionized by the teamsters. "Christ," I said. "That's why that fool Praeger called a special faculty meeting for this afternoon. I thought that it might have something to do with the students, so I was planning to go home. We must begin rounding up the troops."

I went over the list mentally. If only Albert Backen had got tenure, he could have been counted on. Dear Albert—he is now teaching symbolic logic at the College of Petroleum and Minerals in Saudi Arabia. We receive the odd Polaroid showing him riding a camel. But then there was old Lash, who, when and if he was sober, would surely understand the delicacy of the situation. Oddly enough, there was even Feist. Bleibermacher once made a pass at Edwina Gwan, and Feist has been waiting for his chance ever since. It occurred to me that, at the moment, the most constructive thing I could do was to find Lash and explain why his vote was needed. No doubt he would be in the bar at the Faculty Club.

Indeed, I readily located Lash perched on his favorite stool, within easy reach of a wine glass. He was regaling the bartender with some tale that seemed to focus on himself, Wittgenstein and three stoats when I interrupted and tried to explain the impending crisis. "You say, George," he said with apparent anxiety, "Lang has made a fortune in clowns? Does he own a circus?"

I could see that I was fighting a hopeless battle, and after a decent interval I took my leave so as not to miss any of the meeting.

The room was packed. Bleibermacher and Praeger were sitting next to each other in the front row and talking amiably. Bleibermacher's hair, newly transplanted, had the burnished sheen of a vervet's tail. He was wearing a mauve foulard, no doubt woven by Lang's microbes. Praeger showed all the false bonhomie of a television weathercaster as his eyes wandered over the room, counting votes. He stood up and called the meeting to order.

The first piece of business, it turned out, was a proposal to merge the astrophysics and economics departments. A joint degree in "astronomics" would be offered, thus, in Praeger's words, "unifying two areas of fecund speculation." The proposal passed unanimously. The next item was pure Feist. It seemed that Edwina Gwan had applied to our place to do graduate work. The only difficulty was her academic report, which, when plotted, had the wild inconsistency of a graph showing the changing price of shares in a small semiconductor company. She had, needless to say, A's in the several courses she had taken with Feist. Bleibermacher had given her one A followed by three D's. Feist introduced a special petition that would allow her to continue her "important work" as his graduate assistant.

Bleibermacher made an impassioned speech in which he compared the Apollonians and the Dionysians. His voice rose as he thundered "Let no one ignorant of geometry enter here!"—not immediately applicable to Gwan who is studying comparative literature. But as he sat down there was a resounding volley of applause led by Praeger, who commented on the importance of maintaining academic standards "against a rising tide of materialism." This carried the day, and Feist's proposal was narrowly defeated. Alice Dodd passed me a note that read "After many a semester dies the Gwan." Then we turned to the main act—Lang's degree.

First, Praeger made an endless speech. For some reason, he pronounced the word "clone" as if it had two syllables: "clo-un." He said that he saw in Lang a sort of modern Henry Ford, and he added that Lang had a new strain of bugs, which produced a substance that bore an eerie resemblance to stainless steel. At every mention of Lang, Praeger almost visibly salivated. Bleibermacher looked like a giant cat swimming in a pool of cream. His contented smile became positively Buddhic. In front of him were endless years of comparing things on a salary greater than even that of the football coach.

Praeger sat down to only scattered applause, which indicated that the vote might be close. Feist stood up and delivered a bitter speech arguing that Lang should first obtain his high-school equivalency diploma. He and Bleibermacher exchanged glances that could have bored holes in a diamond. I thoroughly enjoyed every minute of it. Then Praeger called for a vote. Hands were raised, and the registrar took a count. It was a tie, which meant that Praeger would get the chance to cast the deciding vote. But just as he was about to do so, in staggered Lash, his face the color of the warning lights on an ambulance. He had to be helped to his chair, and no sooner was he seated than he stood up and demanded a reading of the minutes of the last meeting. The situation was explained to him, and he said in a loud, clear voice "I will have none of it." There was a stunned silence and we all filed out. As I passed him, Lash said to me, "I never liked the circus much—not even when I was a boy."

It took some time for the dust to settle. Edwina Gwan is now our department secretary. Nafken has gone to Harvard Medical School. The Herschel J. Lang Chair in Cognitive Studies, with Bleibermacher firmly attached to it, was moved bodily to a large university in Texas. Bleibermacher sent me a photograph of the ceremony at which Lang was finally awarded his honorary degree. With his right hand Lang is accepting his diploma. With his left hand he is slipping the president an envelope. He has the look of someone who has finally arrived.

───────

Some weeks after this story was published in The New Yorker, *I received a letter from Saudi Arabia. It was from an American who was actually teaching at the College of Petroleum and Minerals. It was a friendly letter informing me that he had found my story quite funny, even the part about the fate of poor Albert Backen. But he also enclosed a Polaroid. It was, I gathered, a photograph of the interior of his living room there. It seemed to have a curved roof like a Quonset hut. But every square inch of the walls were papered with old* New Yorker *covers. He must have loved the magazine. I showed the photograph to William Shawn. He was as impressed as I was.*

Index